DATE DUE

FEB 12 2010	
OCT 30 2010	
DEC 08 2010	
FEB 17 2012	

Cell Biology for Biotechnologists

Cell Biology for Biotechnologists

Shaleesha A. Stanley

Alpha Science International Ltd.
Oxford, U.K.

Cell Biology for Biotechnologists
182 pgs | 15 tbs | 57 figs

Shaleesha A. Stanley
Department of Biotechnology
Jeppiar Engineering College
Jeppiar Nagar, Chennai
Tamil Nadu, India

ALPHA SCIENCE INTERNATIONAL LTD.

7200 The Quorum, Oxford Business Park North
Garsington Road, Oxford OX4 2JZ, U.K.

www.alphasci.com

ISBN 978-1-84265-393-7

Printed in India

Preface

The challenge in writing a textbook is to zero in on the concepts that are most important for the understanding of the subject to be dealt with. Recognizing that it is impossible to be comprehensive, I endeavored to ensure that this book should provide a survey of cell biology that students can easily read and digest.

More than 50 years ago, E.B. Wilson wrote, "The key to every biological problem must finally be sought in the cell". Yet, until very recently, cell biology is taught to biology majors as a specialized course leaving out many essential topics such as endocytosis, chemotaxis, cell movement, cell adhesion etc. Recently, most of the Universities have refined the course curriculum and have paid attention to all the topics thereby adding value to the syllabi.

This textbook has been organized into seven chapters each representing a different theme. I have attempted to include structure of cell membranes and their role in maintaining the architecture of the living being. Information on molecular aspects of the cell is added in sufficient depth and the instrumentation part related to cell biology is appended to provide practical insights to cell biology.

Acknowledgements and thanks must go to many friends who have made contributions to the book in several ways. I am most grateful to Dr. Jeppiaar, Chairman of Jeppiaar Educational Trust and Mr. Wilson and Mrs. Regeena Wilson, Directors of Jeppiaar Engineering College for encouragement and constant follow-up towards this book. I place on record my thanks to Narosa Publishing House for their interest in publishing this book.

Finally, I would like to thank my husband, Dr. V. Amalan Stanley for editing and reviewing this book with much patience and my daughters, Carol and Hepsi, to whom I could not devote precious time while preparing this book.

Constructive suggestions and criticisms are welcome.

Shaleesha A. Stanley
shaleesha.stanley@gmail.com

Contents

CHAPTER 1

Introduction to the Cell

1.1 EVOLUTION AND HISTORY OF THE CELL

The biological science that deals with the study of structure, function, molecular organization, growth, reproduction and genetics of cells is broadly termed as cytology (Gr., Kytos – hollow vessel or cell; logous – to discourse) or cell biology. Cell biology or cytology is devoted to the study of structures and functions of typical and specialized cells. The results of these studies are used to formulate the generalization applied to almost all cells as well as to provide the basic understanding of how a particular cell type carries out its specific functions. **Cell is the structural and functional unit of all basic living organisms.**

Terminology

Robert Hooke, in 1665 observed a piece of cork under a primitive microscope and recorded cells in the form of cavities with cellulose walls and in the year 1674, Leeuwenhoek, confirmed these structures. This structural unit is now known as the unit of life and the concept that the cell is the basic structural and functional unit of life is known as the cell theory. Two German, scientists, Matthias Schleiden and Theoder Schwann formulated cell theory in the year 1838 – 39. Later, in 1858, Louis Pasteur, a French scientist gave experimental evidences to the cell theory and put forth two main components of the cell, namely, (a) all living things are composed of cells and (b) all the existing cells arise from pre-existing cells.

Ancient Greek philosophers such as Aristotle (384 – 322 B.C.) and Paracelsus concluded, "All animals and plants, however, complicated are constituted of a few elements which are repeated in each of them".

Exceptions to the Cell Theory

Organisms that do not fit into the cell theory by its physical, chemical and biological characteristic features are illustrated in **Fig. 1.1.**

1. Viruses

- Are simple in structure
- Lack in internal organization
- Infectious
- Lack in internal organization

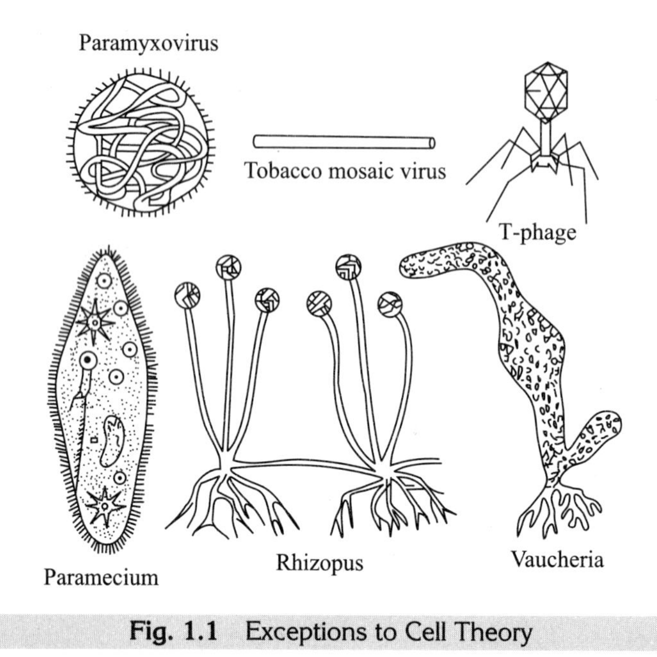

Fig. 1.1 Exceptions to Cell Theory

- Ultramicroscopic particle
- Multiply within the host cell
- Grouped as primitive organisms

2. **Mycoplasma, Viroids**
3. **Some Organisms in the Protozoa Group** such as paramecium, fungus – Rhizopus, algae – Vaucheria – though all these organisms have an outer frame called the cell wall, they lack internal organs as that of a cell.

Size of the Cell

- The diameter of the cell size varies greatly in plants and animals.
- Smallest cells are found in bacteria (0.2 – 0.5μ)
- Largest cell is the egg of an ostrich measuring 6 inches in diameter around the outside and 3 inches when the shell is removed.
- The largest cell, therefore, is about 75,000 times bigger than the smallest bacterial cell.
- The bacteriophages (or) virsus are still smaller in size but they are not considered to be cellular in their organization.
- Cell size depends on lot of parameters mainly on the number of chromosomes and this is referred as the law of constant volume.
- As the cell size increases, an increase in the volume of the cell is much more than the area of the cell surface.
- The volume increases as the cube of radius while surface area increases as square of radius

- Due to disproportionate increase in cell volume and surface area, adequate area, exchange of material may not be available.
- Metabolically active cells tend to be smaller in size.

Cell Number

- Varies from single cell in a unicellular organism to 60,000 billion cells in a human being who weighs approximately 80 kilogram.
- In some organisms, like blue green algae – pandorina, the cell number is constant (8, 16,1 32 or 54 cells)

1.2 CLASSIFICATION OF THE CELL

Cells are basically classified into two types, namely, prokaryotic and eukaryotic. In prokaryotic cells, no organized nuclei are present and in eukaryotic cells, organized nuclei are present.

Prokaryotes: example – bacteria and blue green algae

Eukaryotes: example – higher plants and animals.

In between, there is another group of organisms that was discovered in 1977 and were called 'Archae' (Archaebacteria). After the discovery of this new organism, the living organisms were grouped into three categories, namely,

a. Bacteria or prokarya
b. Archaea
c. Eukarya

Archaeon

During an expedition to deep sea in the year 1982, from a hot spot of 3 kilometers beneath the Pacific Ocean, a methane producing organism was discovered and it was named after the discoverer Holger Jannasch as *Methanococcus jannaschii*. The nucleotides were fully sequenced and compared with the sequences of genome of bacteria and eukaryotes. Among the genes, 50 percent were new to the field of science and it was found that it lives at temperature ranging from 48°C – 94°C and pressures of more than 200 atmospheres. Some species of Methanococcus grow in brine, 10 times higher than seawater. Oxygen is harmful and it's an autotrophy utilizing carbon dioxide, nitrogen and hydrogen.

1.3 STRUCTURE OF THE CELL

The structure of the cell is divided into three components namely plasma membrane, cytoplasm and the nucleus.

The plasma membrane is the thin outer limiting membrane of the cell. It provides mechanical support and external shape to the protoplasm. It also checks the entry and exit of substances from and into the cells (permeability). This membrane maintains the osmotic equilibrium between the cytoplasm of the cell and the surrounding. The plasma membrane also ingests large sized foreign food particles (endocytosis) and the process of passing the materials outside the cell (exocytosis).

The mass of protoplasm (Fig. 1.2) lying outside the nucleus is called the cytoplasm. It is a colorless, homogenous, translucent, colloidal fluid. It consists of various molecules such as water, salts of Na, K and other metals and organic compounds such as carbohydrates, lipids, proteins, nucleoproteins, nucleic acids and enzymes.

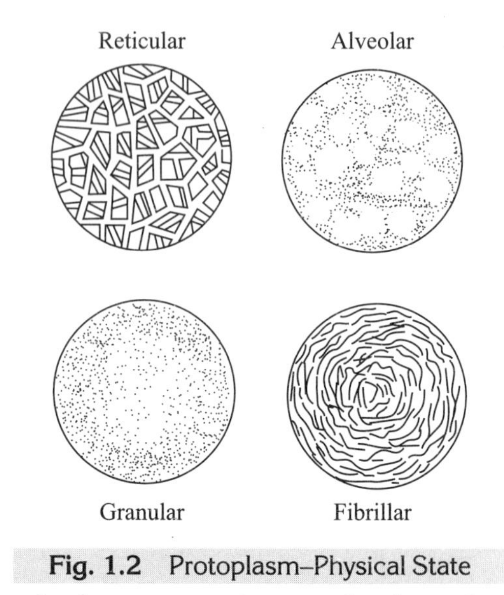

Fig. 1.2 Protoplasm–Physical State

The peripheral part of cytoplasm is non granular and clear and is called as ectoplasm. The inner portion is granular and less viscous known as endoplasm.

The homogenous ground substance of the cytoplasm is known as hyaloplasm or kinoplasm. Inside the hyaloplasm are suspended substances and structures called the cell inclusions or organelles. Cell inclusions or organelles are organized structures of cytoplasm capable of growth and in some cases multiplication and they do specific functions. The cell organelles are endoplasmic reticulum, ribosome's, mitochondria, Golgi complex, lysosomes, centrosomes, basal granules, tonofibrils, plastids etc.

Nucleus is the controlling part of the cell and is first observed by Robert Brown in 1831. Nucleus is present in almost all animal and plant cells except mature erythrocytes of mammals (Fig.1.3).

The nucleus occupies of about 10 to 15 per cent of the cell volume in most of the cells. The protoplasm of the nucleus is called karyoplasms or nucleoplasm and it is covered by a double-layered nuclear membrane. The nuclear membrane at intervals has minute pores and this is concerned with the regulation of the flow of materials from the surrounding cytoplasm and nucleoplasm. The nucleus contains a network of chromatin and during cell division the chromatin network is condensed to form chromosomes. The

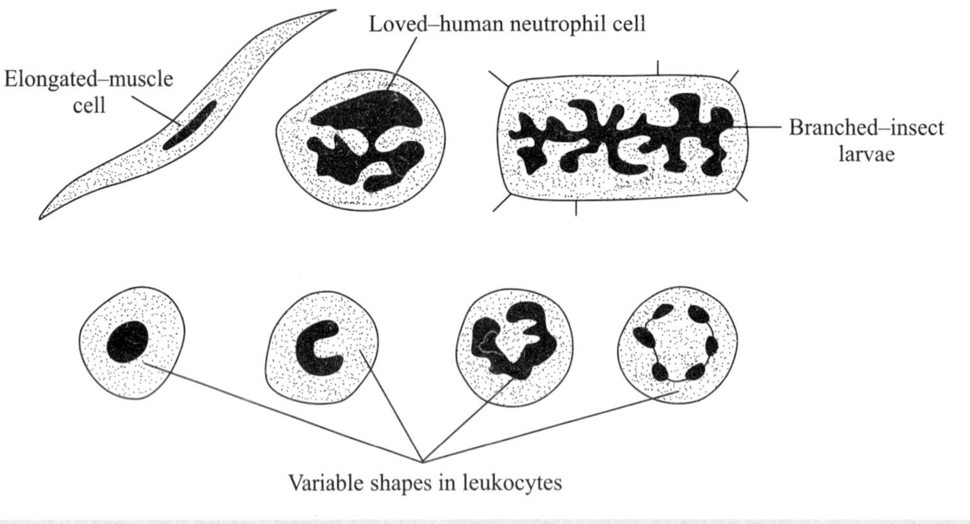

Fig. 1.3 Different Shapes of the Nucleus in Animal Cells

chromatin is composed of two distinct parts, namely, euchromatin and heterochromatin. During interphase and mitosis, the heterochromatin stains darker than euchromatin with Fuelgen and this phenomenon is known as heterpycnosis or differential staining.

The nucleus contains a network of chromatin and during cell division the chromatin network is condensed to form chromosomes. The chromatin is composed of two distinct parts namely, euchromatin and heterochromatin. During interphase and mitosis, the heterochromatin stains darker than euchromatin with Fuelgen and this phenomenon is known as heterpycnosis or differential staining. The nucleoplasm has one or small bodies known as nucleoli discovered first by Fontana in 1871. Limiting membrane does not cover the nucleoli and they are very rich in RNA. The chromatin is rich in deoxy ribonucleic acid (DNA). The nucleus is a dynamic center, which controls and regulates the various metabolic activities of the cell. A cell cannot survive without a nucleus. It acts as a vehicle for the transmission of hereditary characters through chromosomes during cell division.

1.4 SHAPE AND SIZE OF THE CELL

Cell shape varies in different organisms and some organisms like amoeba and leukocytes keep changing their shapes frequently. Some typical shapes of the cells of bacteria are rods, spirals and comas. Acetabularia is a single celled algae with a stack and cap making up 10 cm in height.

The shape of the cell depends upon the function it has to perform. For example, cells like glandular hairs on the leaf, Guard cells of stomata and root hair cells. Stability of the cell shape is provided by the cytoskeleton associated with the inner surface of the plasma membrane of the cell. This is described as the membrane skeleton.

The eukaryotic cells are larger than the prokaryotic cells and they range from 10 to 100 μm. The sizes of the cells of the unicellular organisms are larger than the typical multicellular organisms. For example, Amoeba proteus is biggest among the unicellular organisms in length (1000 μm). Largest animal cell, the egg of ostrich ranges in diameter (18 cm) and some nerve cells of human beings have a long tail called axons (Fig. 1.4).

Though the basic shape of the eukaryotic cell is spherical, the shape is ultimately determined by the specific function of the cell. Thus, the shapes of the cells are varied or can be constant and there can be variable or irregular shape occurring amoeba and white blood cells or leukocytes. The shapes are constant for protests such as paramecium, euglena, plants and animals. In unicellular organisms, the shape of the cell is maintained by the tough plasma membrane and exoskeleton. In a multicellular organism, the shape of the cell depends mainly on the functional adaptations and partly on the surface tension, viscosity of the protoplasm, cytoskeleton of microtubules, microfilaments and intermediary filaments and the mechanical action exerted by the neighboring cells and finally by the rigidity of the plasma membrane.

The shape of the cell varies from organ to organ and forms animal to animal and sometimes, the cells of same organ may also vary in shapes. For example, the squamous epithelial cells have diverse shapes such as polyhedral, flattened, cuboidal, discoidal, spherical etc. The smooth muscle cells are spindle shaped and nerve cells or neurons are elongated and the chromatophores or pigment cells of skin are branched. In plants, the shape of the cell depends on the specific function of the cell. For example, glandular hair cells on a leaf, guard cells of stomata and root hair cells have special shape.

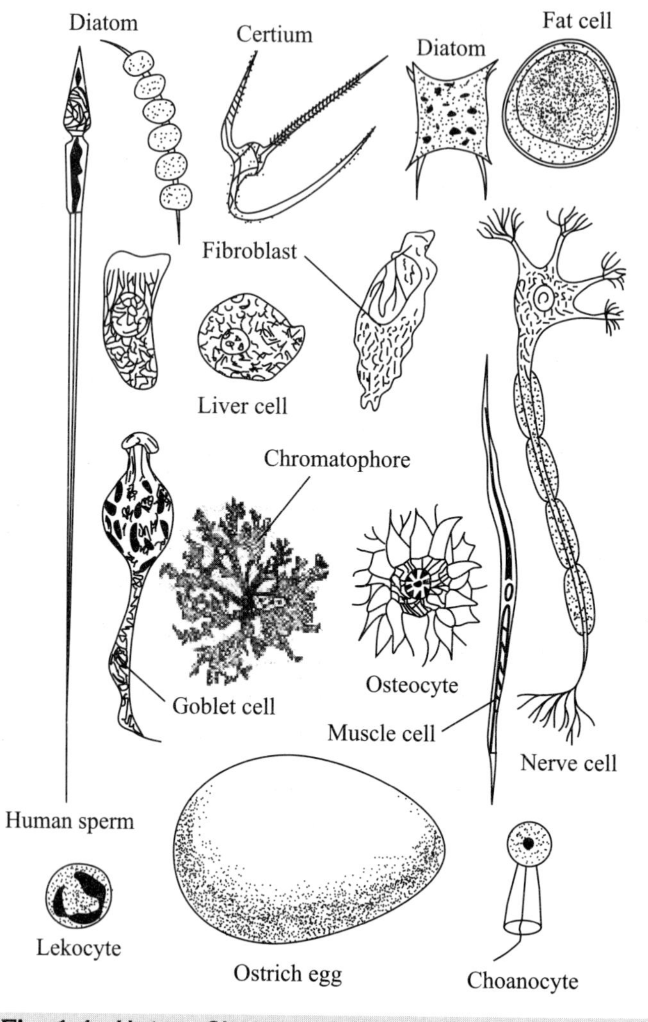

Fig. 1.4 Various Shapes and Sizes of Eukaryotic Cells

1.5 OBSERVATION OF CELLS – MICROSCOPY

Microscopes are instruments, most characteristic of cell biology laboratory. The magnification it provides enables us to see microorganisms and their structures otherwise invisible to the naked eye. The magnification attainable by the microscopes ranges from x100 to x400, 000. In addition, several different kinds of microscopy are available, and many techniques have been developed by which specimens of microorganisms can be prepared for examination. Each type of microscopy and each method of preparing specimens for examination offer advantages for demonstration of specific morphological features. This part describes the categories of microscopes, particularly, bright field and dark field microscopy, fluorescent microscopy, phase contrast microscopy, transmission electron microscopy, scanning electron microscopy and limitations of electron microscopy. A comparison of different types of microscopy and preparations of specimens for different microscopy are also briefed.

Microscopes and Microscopy

Cells are tiny but complex bodies and it is difficult to see their structures and the more is the difficulty to understand their molecular composition and still difficult to find out the function of their various components. In the search for information about the structure and composition of cells, the cell biologists immediately face two limitations and they are the dimension and the transparent nature of the cells.

The human eye has limited distinguishing or resolving power. The ability of an observational instrument such as the human eye or a microscope to reveal details of structure is expressed in terms of limit of resolution. The unaided human eye under optimal conditions in green light (to which it is most sensitive) cannot distinguish between points less than about 0.1 mm or 100μm apart. Structural details smaller than this, e.g., cell, is unresolved unless some instrument capable of higher resolution is used.

Magnification, the increase in size of optical image over the size of the object being viewed, is of no use unless the observational system can resolve the various parts of the structure being examined. Increased magnification without improved resolution results only in a large blurred image. The human eye has no power of magnification. Therefore, magnifying glasses may be used to magnify images unto 10 times and a simple light microscope can magnify 1500 times.

Microscopes are of two categories, light (or optical) and electron, depending upon the principle on which magnification is based. Light microscopy, in which magnification is obtained by a system of optical lenses using light waves, includes: -

a) Bright-field
b) Dark-field
c) Phase-contrast
d) Transmission electron microscope
e) Scanning electron microscopy

The electron microscope, as the name suggests, uses a beam of electrons in the place of light waves to produce the image. Specimens can be examined by either transmission or scanning electron microscopy. The basic and widely used microscope is the bright-field microscope and other types of microscopy are for special purposes or research investigations.

a) Bright-field microscopy

In bright-field microscopy, the microscopic filed (the area observed) is brightly lighted and the microorganisms appear dark because they absorb some of the light. Ordinarily, microorganisms do not absorb much light, but staining them with a dye greatly increases their light-absorbing ability, resulting in greater contrast and color differentiation. The optical parts of a typical bright field microscope and the path the light rays follow to produce enlargement, or magnification, of the object are shown diagrammatically. Generally, microscopes of this type produce a useful magnification of about x1000 to x2000. At magnifications greater than x2000, the image becomes blurry for reasons explained below.

Resolving power The basic limitation of the bright-field microscope lies not with the magnification issue but with the resolving power, the ability to distinguish two adjacent points as distinct and separate. Mere increase in size (greater magnification) without the ability to distinguish structural details (greater resolution) is not beneficial. To state it differently, the largest magnification produced by a microscope may not be the most useful one because the image obtained may be unclear or blurry. The more lines or dots per unit area that can be seen distinctly as separate lines or dots, the greater is the resolving power of the microscope system. The resolving power of a microscope is a function of the wavelength of light used and the numerical aperture (NA) of the lens system.

Numerical aperture (NA) The angle (θ) sub intended by the optical axis and the outermost rays still covered by the objective is the measure of the aperture of the objective. It is the half aperture angle. The magnitude of this angle is expressed as a sine value. The sine value of the half aperture angle multiplied by the refractive index n of the medium filling the space between the front lens and the cover slip gives the numerical aperture (NA): NA=n sin θ. With dry objectives the value of n is, since 1 is the refractive index of air, when immersion oil is used as the medium, n is 1.56 and if θ is 58°, then

$$NA = n \sin\theta = 1.56 \times \sin 58^{\circ} = 1.56 \times 0.85 = 1.33.$$

The degree to which microscope objectives can be altered to increase the NA is limited. The maximum NA for a dry objective is less than 1.0, the oil immersion objective lens have an NA of slightly greater than 1.0 (1.2 to 1.4). The wavelength of light used in optical microscopes is also limited. The visible light range is between 400 nm (blue light) and 700 nm (red light) or 0.4 μm to 0.7 μm. (The abbreviation nm stands for nanometer and is equal to 0.001 4 μm or 10^{-9}m. Thus it is apparent that the resolving power of the optical microscope is restricted by the limiting values of the NA and the wavelength of the visible light.

Limit of resolution The limit of resolution is the smallest distance by which two objects can be separated and still is distinguishable as two separate objects. The greatest resolution in light microscopy is obtained with the shortest wavelength of visible light and an objective with the maximum NA. The relationship between NA and resolution can be expressed as follows: -

$$d = \lambda/2\ NA$$

Where d = resolution and λ = wavelength of light. Using the values 1.3 for NA and 0.55 μm, the wavelength of green light, for λ, resolution can be calculated as d = 0.55 / 2 x 1.30 = 0.21 μm. From these calculations we may conclude that the smallest details that can be seen by the typical light microscopes are those having dimensions of approximately 0.2 μm.

Magnification Magnification beyond the resolving power is of no value since the larger image will be less distinct in detail and fuzzy in appearance. The situation is analogous to that of a movie screen. If we move closer to the screen the image is larger and is also less sharp than when viewed at a distance. Most laboratory microscopes are equipped with three objectives, each capable of a different degree of magnification. These are referred to as the oil-immersion, high dry and low power objectives. The primary magnification provided by each objective is engraved ion its barrel. The total magnification of the system is determined by multiplying the magnifying power of the objective by that of the eyepiece. Generally, an eyepiece having a magnification of x10 is used, although eyepieces of higher or lower magnifications are available.

b) Dark field microscopy

The effect produced by the dark-field technique is that of a dark background against which objects are brilliantly illuminated. This is accomplished by the equipping the light microscope with a special kind of condenser that transmits a hollow cone of light from the source of illumination. Most of the light directed through condenser does not enter the objective and the field is essentially dark. However, some of the light rays will be scattered (diffracted) if the transparent medium contains objects such as microbial cells. This diffracted light will enter the objective and reach the eye and thus the object or microbial cell in this case will appear bright in an otherwise dark microscopic field. Dark-field microscopy is particularly valuable for the examination of unstained microorganisms suspended in fluid – wet mount and hanging drop preparations.

Many chemical substances absorb light. After absorbing light of a particular wavelength and energy, some substances will then emit light of a longer wavelength and a lesser energy content. Such substances are called fluorescent and this phenomenon is called fluorescence. Application of this phenomenon is the basis of fluorescence microscopy. In practice, microorganisms are stained with a fluorescence dye and then illuminated with blue light and the blue light is absorbed and green light emitted by the dye. The special features of fluorescence microscopy with the respect to illumination of the specimen are that it functions as the exciter filter to remove all but blue light. The barrier filter blocks out blue light and allows green light to pass through and reach the eye. Barrier filters are selected on the basis of the dye used.

c) Phase contrast microscopy

Phase contrast microscopy is extremely valuable for studying living unstained cells and is widely used in applied and theoretical biological studies. It uses a conventional light microscope fitted with a phase-contrast objective and a phase contrast condenser. This special optical system makes it possible to distinguish unstained structures within a cell, which differ only slightly in their refractive indices or thick nesses.

In principle, this technique is based on the fact that light passing through one material and into another material of a slightly different refractive index and / or thickness will undergo a change in phase. These differences in phase, or wave front irregularities, are translated into variations in brightness of the structures and hence are detectable by the eye. With phase-contrast microscopy, it is possible to reveal differences in cells and their structures not discernible by other microscopic methods.

d) Transmission electron microscopy

Electron microscopy differs markedly and in many respects from the optical microscopic techniques. The electron microscope provides tremendous useful magnification, because of the much higher resolution obtainable with the extremely short wavelength of the electron beam used to magnify the specimen. The electron microscopes use electron beams and magnetic fields to produce the image whereas the light microscopes uses light waves and glass lenses.

For electron microscopy, the specimen to be examined is prepared as an extremely thin dry film on small screens and is introduced into the instrument at a point between the magnetic condenser and the magnetic objective. This point is comparable to the stage of the light microscope. The magnified image may be viewed on a fluorescent screen through an airtight 'window' or recorded on a photographic plate by a camera built into the instrument. Numerous techniques are available for use with electron microscopy, which extend its usefulness in characterizing cellular structure. Some of these are described below.

Shadow casting This technique involves depositing an extremely thin layer of metal at an oblique angle on the organism so that the organism produces a shadow on the uncoated side. The shadowing technique produces a topographical representation of the surface of the specimen.

Negative staining An electron-dense material such as phosphotungstic acid can be used as a 'stain' to outline the object. The electron-opaque phosphotungstate does not penetrate structures but forms thick deposits in crevices. Fine details of objects such as viruses, bacterial flagella can be viewed through this technique.

Ultra thin sectioning In order to make observations of intracellular structures, the material for examination must be extremely thin. An intact microbial cell is too thick to allow distinct visualization of its internal fine structure by electron microscopy. However, techniques are available for sectioning (slicing) a bacterial cell. For example, bacterial cells can be embedded in a plastic material and then this 'block' can be cut into ultrathin slices as thin as 60 nm. These slices are then prepared for microscopic examinations. As you might expect, the slices will reveal cells sliced at different levels and at different angles. Improvement

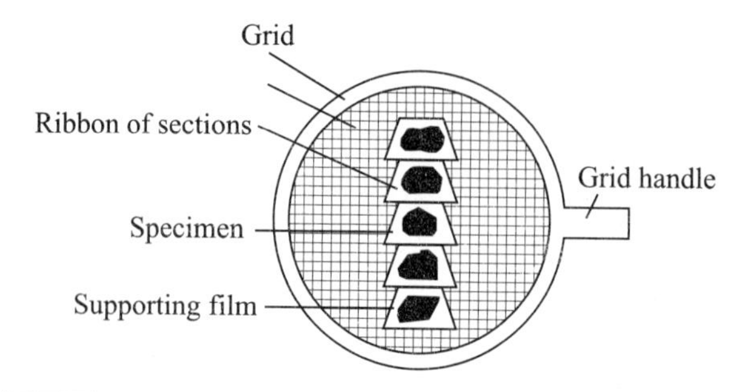

Fig. 1.5 Ultra Thin Sections Placed on a Grid (After Sheeler and Bianchi, 1987)

in contrast of structures is possible through use of special electron microscope stains such as uranium and lanthanum salts.

Freeze-etching Freeze-etching was developed to prepare sections of the specimen without resorting to the chemical treatment of the fixation process, which can produce artifacts. The specimen is sectioned while contained in a frozen block. Carbon replicas of these exposed surfaces are then prepared to reveal internal structures of the cells.

Autoradiography Autoradiography is a cytochemical method in which the location of a particular chemical constituent in a specimen is determined by observing the site at which radioactive material becomes positioned. The cells are first exposed to the radioactive substance to permit its uptake. In practice, the specimen prepared for the microscopic examination is covered with a layer of photographic emulsion and stored in the dark for a period of time. The ionizing radiation emitted during the decay of a radioactive substance produces latent images in the emulsion, and after photographic processing, the developed image is seen as grains of silver in the preparation.

e) Scanning electron microscopy

In scanning electron microscopy, the specimen is subjected to a narrow electron beam, which rapidly moves over (scans) the surface of the specimen. This causes the release of a shower of secondary electrons and other types of radiation from the specimen surface. The intensity of these secondary electrons depends on the shape and the chemical composition of the irradiated object. The secondary electrons are collected by a detector, which generates an electronic signal. These signals are then scanned in the manner of a television system to produce an image on a cathode ray tube. The scanning electron microscopy lacks the resolving power obtainable with the transmission electron microscope but has the advantage of revealing a striking three-dimensional picture. The surface topography of a specimen can be revealed with clarity and a depth of field not possible by any other method.

Limitations of Electron Microscopy

Despite the great advantage of tremendous resolution and magnification, there are several limitations to electron microscopy. For example, the specimen being examined is in a chamber that is under a very high vacuum. Thus cells cannot be examined in a living state. In addition, the drying process may alter some morphological characteristics. Another limitation of the technique is the low penetration power of the electron beam, necessitating the use of thin sections to reveal the internal structures of the cell.

The real problem confronting the researcher who attempts to unravel the fine intracellular structure of the microbial cell is identification of intracellular material. Frequently, it is necessary to correlate results obtained with the same organism viewed by different microscopic techniques, example, phase-contrast, bright field (stained preparations) and electron microscopy. Each method contributes different kinds of information. Interpretation of this information, particularly comparison of what is revealed by each technique, makes it possible to identify cellular structures. But considerable experience in microscopy is required before a researcher can correctly interpret the results.

Table 1 Comparison of Different Types of Microscopy

S.No	*Type of Microscopy*	*Maximum magnification*	*Appearance of specimen*	*Useful Applications*
1.	Bright-field	1,000-2,000	Stained or unstained.	Gross morphological features of bacteria, yeasts, molds, algae and protozoa
2.	Dark-field	1,000-2,000	Unstained in a dark background	For microorganisms that exhibit some characteristic morphological feature in the living state and in fluid suspension.
3.	Fluorescence	1,000-2,000	Bright and colored with a fluorescent dye	Diagnostic techniques where fluorescent dye fixed to organism reveals the organism's identity
4.	Phase contrast	1,000-2,000	Varying degrees of darkness	Examination of cellular structures in living cells of the larger micro-organisms. Example – yeasts, algae, few bacteria and some protozoa
5.	Electron	200,000-400,000	Viewed on fluorescent screen	Examination of viruses and ultra structure of microbial cells.

Preparation of Specimens for Observation Under Light Microscopes

Two general techniques are used to prepare specimens for light microscopic examination. One is to suspend organisms in a liquid (the wet mount or the hanging drop techniques) and the other is to dry, fix and stain films or smears of the specimen.

The Wet-Mount and Hanging Drop Technique

Wet preparations permit examination of organisms in a normal living condition. A wet mount is made by placing a drop of fluid containing the organisms onto a glass slide and covering the drop with a cover slip. To reduce the rate of evaporation and exclude the effect of air currents, the drop may be ringed with petroleum jelly or a similar material to provide a seal between the slide and the cover slip. A special slide with a circular concave depression is sometimes used for examination of wet preparations. A suspension of microbial specimen is placed on a cover slip, and then inverted over the concave depression to produce a 'hanging drop' of the specimen.

Fixed, Stanned Smears

Fixed, stained preparations are most frequently used for the observation or the morphological characteristics of bacteria. The advantages of this procedure are that (a) the cells are made more clearly visible after they are colored and (b) differences between cells of different species and within the same species can be demonstrated by use of appropriate staining solutions (differential or selective staining). The essential steps in the preparation of a fixed, stained smear are (a) preparation of a film or smear (b) fixation and (c) application of one or more staining solutions.

Stains

Based on the molecular complexity, the stains are classified into groups such as triphenylmethane dyes, oxazine dyes and thiazine dyes. According to the chemical behavior, the stains are again grouped into acidic, basic and neutral ones.

Preparation of Specimens for Observation Under Transmission Electron Microscopy

The routine procedure for preparing specimen for transmission electron microscope entails fixation, dehydration, staining and sectioning, all of that are similar to light microscopy. However, the most significant difference is the need to prepare thin (ultra) sections. The following are the steps in electron microscopy: -

Fixation: Osmium tetroxide, potassium permanganate, formalin, glutaraldehyde

Dehydration: Increasing concentration of ethanol (or acetone) followed by propylene oxide

Embedding: Araldite : vestoplaw, Epan 812, Maraglas : Durcopan

Sectioning: Usually 10-100 nm thick sections cut with a glass or diamond knife on an ultramicrotome

Mounting: On a perforated metal disc (grid) usually covered with formvar or paralodian

Staining: With salts of heavy metals such as lead acetate, lead citrate, lead hydroxide, uranyl acetate and phosphotungistic acid

Viewing: Grid is placed between the condenser and objective lenses in a vacuum and the image is viewed on a phosphorescent screen.

Questions

1. Define a cell.
2. Who discovered the cell? Explain, how is the growth of cell biology linked with the improvement in instrumental analysis?
3. What is cell theory and who put it forward?
4. Explain the exceptions to cell theory.
5. What do you understand by the term? 'Exceptions' to cell theory?

6. What is meant by classical period of cell biology?
7. What do you know about protoplasm theory and organismal theory?
8. What are the branches in cell biology?
9. What is the scope of Cell Biology?
10. Explain about cell size.
11. How will you classify cell and on what basis?
12. Enumerate the uniqueness of Archaeon.
13. What is protoplasm?
14. Sketch any five different shapes of the cell.
15. Classify microscopes based on its magnification concept.
16. Explain: - (a) Bright field microscopy (b) Dark field microscopy (c) phase contrast microscopy (d) Transmission electron microscopy (e) Scanning electron microscopy
17. Define (a) Resolving power (b) Numerical Aperture (NA) (c) Limit of Resolution (d) Magnification
18. Explain Auto radiography.
19. List the limitations of Electron Microscopy.
20. Prepare a comparison statement on the different types of microscopy covering types, magnification and its useful applications.

CHAPTER 2

Cell Structure and Function of Organelles

Present day living cells are classified as prokaryotic (bacteria and associated classes) and eukaryotic. Although they have a relatively simple structure, prokaryotic cells are biochemically versatile and diverse: for example, all of the major metabolic pathways can be found in bacteria, including the three principal energy-yielding processes of glycolysis, respiration and photosynthesis. Eukaryotic cells are larger and more complex than prokaryotic cells and contain more DNA, together with components that allow this DNA to be handled in elaborate ways.

The DNA of the eukaryotic cell is enclosed in a membrane-bounded nucleus, while the cytoplasm contains many other membrane-bounded organelles, including mitochondria, which carry out the oxidation of food molecules, and in plant cells, chloroplasts, which carry out photosynthesis. Mitochondria and chloroplasts are almost certainly the descendants of earlier prokaryotic cells that established themselves as internal symbiots of a larger anaerobic cell. Eukaryotic cells are also unique in containing a cytoskeleton of protein filaments that helps to organize the cytoplasm and provide the machinery for movement.

2.1 PROKARYOTIC CELLS

The prokaryotic cells (Gr., *pro* – primitive or before; *karyon* – nucleus) are small, simple and most primitive. They are probably the first to come into existence perhaps 3.5 billion years ago. For example, the stromatolites (i.e., giant colonies of extinct cyanobacteria or blue green algae) of Western Australia are known to be atleast 3.5 billion years old. The eukaryotic (Gr., *eu* – well; *karyon* – nucleus) cells have evolved from the prokaryotic cells and the first eukaryotic (nucleated) cells may have risen 1.4 billion years ago.

The prokaryotic cells are the most primitive cells from the morphological point of view. They occur in the bacteria (i.e., mycoplasma, bacteria and cyanobacteria or blue green algae). A prokaryotic cell is essentially a single-enveloped system organized in depth. It consists of central nuclear components (viz., DNA molecule, RNA molecules and nuclear proteins) surrounded by cytoplasmic ground substance, enveloped by a plasma membrane.

Membranes separately enclose neither the nuclear apparatus nor the respiratory enzyme systems, although the inner surface of the plasma membrane itself may serve for enzyme attachment. The cytoplasm of a prokaryotic cell lacks in well-defined cytoplasmic organelles such as endoplasmic reticulum, Golgi apparatus, mitochondria, centrioles etc. In a nutshell, the prokaryotic cells are distinguished from the eukaryotic cells primarily on the basis of what they lack, i.e., prokaryotes lack in the nuclear envelope, and any other cytoplasmic membrane. They also do not contain nucleoli, cytoskeleton (microfilaments and microtubules), centrioles and basal bodies.

Bacteria

Bacteria are the smallest, primitive, simple, unicellular, prokaryotic and microscopic organisms. All bacteria are structurally homogenous and their biochemical activity and their ecological niches are extremely diverse (Fig. 2.1). Bacteria occur almost everywhere i.e., in air, water and soil and inside other organisms. They are found in stagnant ponds and ditches, running streams, rivers, lakes, sea water, foods, petroleum oils, manure heaps, sewage, decaying organic matter of all types, on the body surface, in body cavities, and in the internal tracts of man and animals. Bacteria thrive well in warmth but some can survive at very cold temperatures also. A teaspoonful of soil may contain several hundred million of bacteria and they can lead either an autotrophic or heterotrophic mode of existence. The saprophytic bacteria are of great economic significance for man. Some species of bacteria are pathogenic to plants, animals and man.

Bacteria have a high ratio of surface area of volume because of their small size. They show high metabolic rate because they absorb their nutrients directly through cell membranes. They multiply at a rapid rate. In consequence, due to their high metabolic rate and fast rate of multiplication, bacteria produce marked changes in the environment in a short period.

Among the living organisms that have the smallest mass are mycoplasmas, a type of bacteria. These produce infectious diseases in animals including human beings. Mycoplasmas are unicellular, prokaryotic containing a plasma membrane, DNA, RNA and a metabolic machinery to grow and multiply in the absence of other cells. These mycoplasmas can be filtered through bacterial filters and they do not contain cell wall

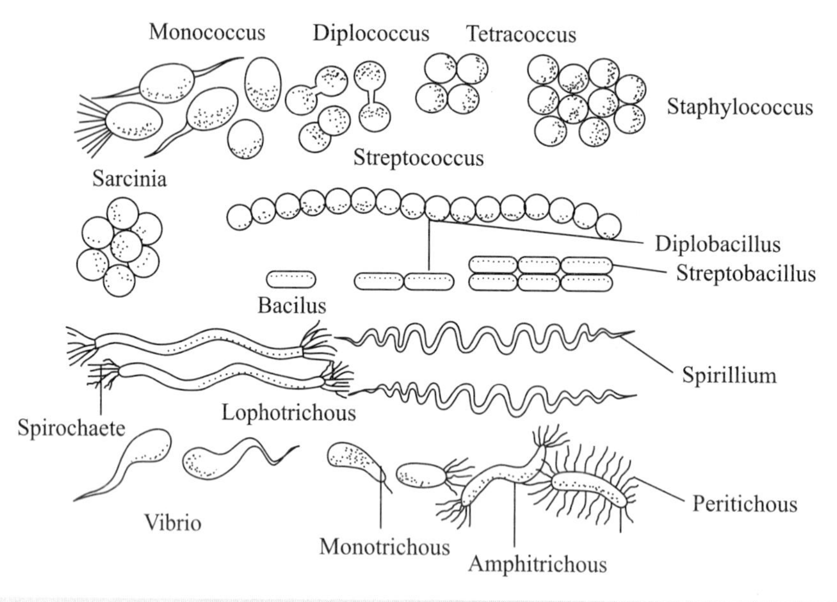

Fig. 2.1 Bacteria – different Forms

and mesosomes. Mycoplasmas are like the viruses and animal cells in having resistance to antibiotics such as penicillin. These mycoplasmas were first discovered by French Scientists, E.Nocard and E.R. Roux in 1898 while studying pleural fluids of cattle suffering from the disease, pleuropneumonia. Similar mycoplasmas were isolated from animals such as sheep, goat, rat, mice and human beings and were named as Pleuropneumonia-like organisms (PPLO) (Fig. 2.2).

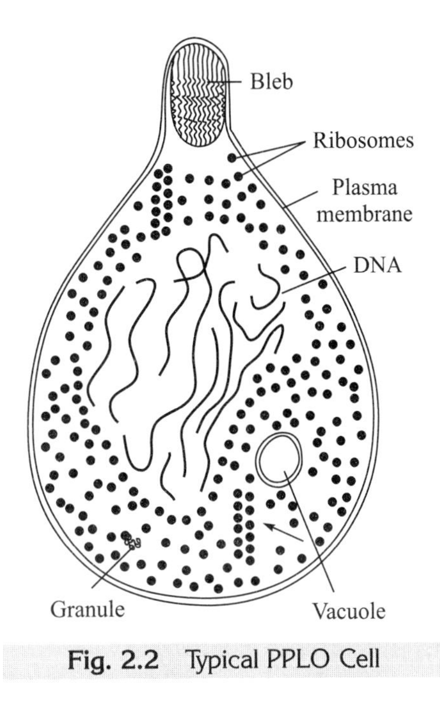

Fig. 2.2 Typical PPLO Cell

Escherichia coli or E.coli is studied as an typical example for a prokaryotic cell. E.coli is a Gram negative, monotrichous, symbiotic bacillus of colon of human beings and other vertebrates. It is heterotrophic, measuring about 2μm long and 1μm wide and it's non-pathogenic producing vitamins such as Vitamin K for human use. Some strains of E.coli are known to recognize and bind specifically to sugar containing target cells on the surface of gut lining of mammals. E.coli is one among the best-studied bacteria in the field of molecular biology and genetic engineering, since it is easy to grow in an artificial medium where it divides every 20 minutes at 37°C under optimal conditions. Thus a single cell becomes 10^9 bacteria in about 20 hours (Fig. 2.3).

Typical structure of E.coli has a plasma membrane. This plasma membrane is a typical fluid mosaic model and external to the plasma membrane occurs a rigid and protective cell wall, which has a complex organization, and it comprises an external membrane. This external membrane is a lipid bilayer traversed by numerous porin channels that allow the diffusion of solutes. Each porin channel is formed by 6 to 8 subunits, each having three suspended hydrocarbon chains. The porin is a polypeptide and it spans the full thickness of outer membrane. The plasma and the external membrane of the cell wall are separated by the periplasmatic space. This space contains a grid or reticulum of petidoglycans and some porin subunits remain attached to the peptidoglycan grid.

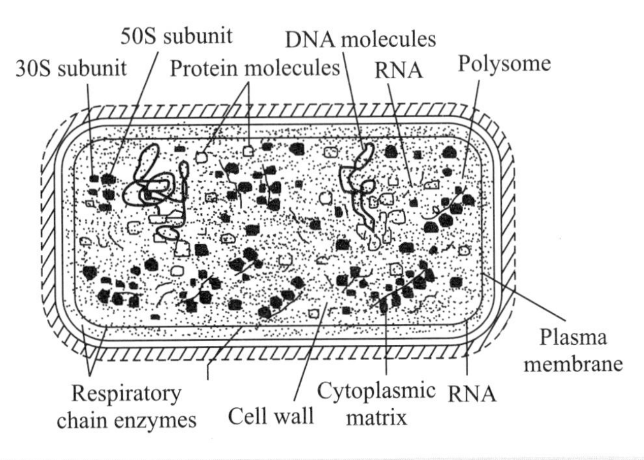

Fig. 2.3 Typical Prokaryotic Cell of Escherichia Coli

The colloidal matrix of E.coli contains about 5000 distinguishable components, ranging from water to DNA. Some of the components are the three types of RNA, enzymes, glycogen, amino acids, monosaccharides and various other molecules. Surrounding the DNA is a dark dense region of matrix containing 20,000 to 30,000 70S type ribosomes, each existing in the form of their two subunits. During protein synthesis, numerous complete ribosomes have read the codes of mRNA molecules to form the polysomes.

2.2 EUKARYOTIC CELLS

The eukaryotic cells (Gr., *eu* – well; *karyon* – nucleus) are two envelope systems and they are very much larger than prokaryotic cells. Secondary membranes envelop the nucleus and other internal organelles. The eukaryotic cells are true cells, which occur in the plants (from algae to angiosperms) and the animals (from protozoa to mammals). The eukaryotic cells have varied shapes, sizes and physiology and the cells are typically composed of plasma membrane, cytoplasm, and organelles such as mitochondria, endoplasmic reticulum, ribosomes, Golgi apparatus and a true nucleus. The nuclear contents such as DNA, RNA, nucleoproteins and nucleolus remain separated from the cytoplasm by a thin perforated nuclear membrane.

Characteristic Features

- Organized nucleus
- Two envelope systems
- Larger than prokaryotes
- Occur in plant and animal cells
- Though they vary in shape, size and number all the cells are typically composed of Plasma membrane, cytoplasm and nucleus and its cytoplasmic organelles such as mitochondria, chloroplast, Golgi complex, endoplasmic reticulum, ribosomes etc
- Nuclear contents viz. chromatin fibres, nucleoplasm, nucleolus remain separated from the cytoplasm by a thin perforated nuclear membrane (Fig. 2.4).

Shape

- Exhibits various shapes and forms
- Typical animal cell is spherical
- Others may be irregular, triangular, tubular, cuboidal, polygonal, oval, cylindrical etc
- Shape vary from animal to animal and also organ to organ
- Shape of the cell is also correlated to the function it perform Ex. Epithelial cells are flat, muscle cells are elongated

External and internal environments also cause variations in the cell due to surface tension, pressure and mechanical stress etc.

Size

- Microscopic

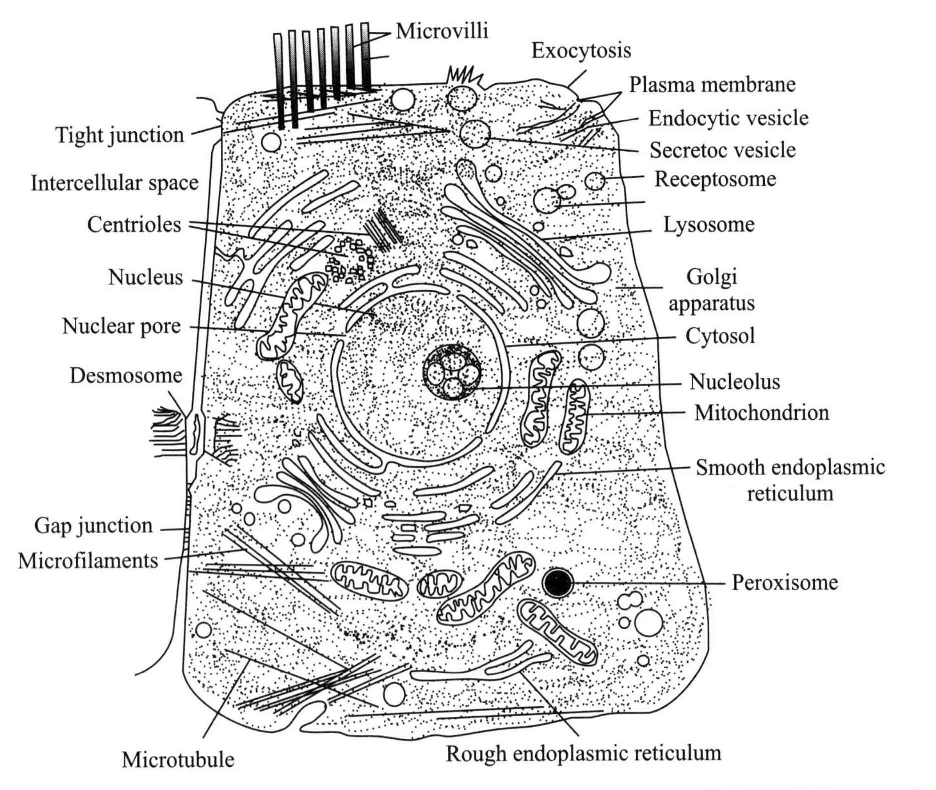

Fig. 2.4 Ultrastructure of a Typical Animal Cell Under Electron Microscope

- Larger than bacterial cells
- Ranges from 1μm - 1,75,000μm
- Largest cell: ostrich egg
- Longest cell – nerve cell (3-3.5 feet long)

Number

- Unicellular (rarely)
- Mostly multicellular
- Smaller organisms have lesser number of cells
- Larger organisms have more number of cells

Typical Structure of an Eukaryotic cell

Main components include

I. Cell wall / plasma membrane
II. Cytoplasm
III. Nucleus

I. Cell wall

- Present only in plant cells (Fig. 2.5)
- semi-rigid, non-living
- External covering of the cell
- Secreted by the cell itself
- Contains a complex polysaccharide carbohydrate called cellulose
- Provides protection and support to the underlying plasma membrane.
- Cell wall has canal like or minute pit like apertures by which the cells remain connected with adjacent cells and they are called plasmodesmata.

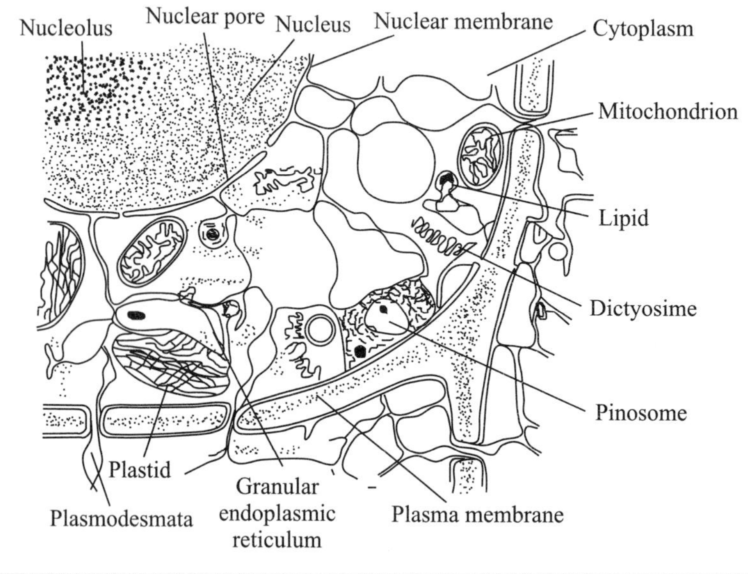

Fig. 2.5 Ultrastructure of a Typical Plant Cell

Plasma membrane

- Plants and animals have
- Called as plasmalemma or cell membrane or plasma membrane
- Living, thin, elastic, porous, semi-permeable
- Contains inner, outer and middle layer of lipids
- Contains pores through which exchange of molecules occur
- FN – provides mechanical support and external form to the protoplasm (cytoplasm and nucleus) and checks the entry and exit of desirable substances into and from the cells

II. Cytoplasm

Cytoplasm will be classified for the convenience of study as

A. Cytoplasmic Matrix
B. Cytoplasmic structures or organelles

A. Cytoplasmic matrix

- Is the space between the plasma membrane and nucleus
- It's an amorphous translucent, colloidal liquid called cytoplasmic matrix or hyaloplasm
- Contains inorganic molecules such as water, salts of Na, K, other metals etc
- Contains organic compounds such as carbohydrates, proteins, nucleoproteins, nucleic acid, DNA, RNA and a variety of enzymes
- Peripheral layer of cytoplasmic matrix is relatively non-granular, viscous, clear and rigid called plasma gel cortex or ectoplasm
- The inner portion is granular, less viscous and is called endoplasm

B. Cytoplasmic structures or organelles Are living membrane bound organelles performing different functions such as respiration, biosynthesis, transportation, support, storage and reproduction.

Some of the important organelles are as follows: -

1. Microtubules
2. Cytoplasmic filaments
3. Centrosome
4. Basal granules or Kinetosomes
5. Cilia and flagella
6. Endoplasmic reticulum
7. Golgi complex
8. Lysosomes
9. Cytoplasmic vacuoles
10. Microbodies
11. Ribosomes
12. Mitochondria
13. Plastids
14. Chloroplasts
15. Endosomes

1. Microtubules Microtubules are numerous ultra fine tubules made up of a protein called tubulin and these structures are present in the cytoplasm of plant and animal cells.

Microtubules comprise 13 individual protofilaments arranged in the form a hollow cylinder.

Function

- Transportation of water ions, small molecules, cytoplasmic streaming
- Formation of fibres or asters of the mitotic and meiotic spindle during cell division
- Form the structural units of the centrioles, basal granules, cilia and flagella.

2. Cytoplasmic filaments Cytoplasmic filaments are ultra fine, proteinaceous and solid filaments of various sizes.

The three recognized sizes are (a) Microfilaments – smallest, 40 – 60°A in diameter and it occurs next to plasma membrane, (b) Myosin filaments – found first in muscle cells and they contain a protein called actin and these get involved in the cell movement and (c) filaments – 100A° (10nm)

Function Helps in the movement of the cell itself and the materials within the cell.

3. Centrosomes

- Centrosome is the dense cytoplasm found near the nucleus of the animal cells
- During cell division, centrosomes contain two rod shaped granules known as centrioles
- Centrioles consist of nine fibrillar units and each fibrillar unit is found to contain three micro-tubules

Function During cell division, centrioles form the spindle of microtubules, which helps in the separation, and movement of chromosomes.

4. Basal Granules (or) Kinetosomes

- The animal and plant cells that have locomotory organelles such as flagella and cilia contain spherical bodies known as basal granules or kinetosomes at the base of the flagella or cilia.
- The basal granules or kinetosomes remain embedded in the ectoplasm and are composed of nine fibrils
- Each fibril consists of three microtubules, out of which two enter into the cilia or flagella.
- The basal granules may have DNA or RNA.

5. Cilia & Flagella

- The cells of many unicellular organisms and ciliated epithelium of multicellular organisms consist of some hair like cytoplasmic projections outside the surface of the cell. These are known as cilia or flagella.
- Cilia are typically 2 - 10μm long and 0.5μm in diameter and are numerous on those cells that have cilia
- Flagella are longer (100-200μm long), and have the same diameter and structure as that of cilia but are usually one or two per cell.
- Both structures are composed of microtubular framework enclosed by an extension of the plasma membrane.
- Both originate from the basal granules and chemically consist of tubulin and dynein proteins and ATP, Adenosine triphosphate.

Function

- Both are motile appendages
- Movement is achieved by sliding microtubules
- Cells lining the air passages in the lungs
- Cilia- wave like motion sweeping minute particles into the lungs
- Flagella – propel the cell itself.

6. Endoplasmic Reticulum

- The cytoplasmic matrix consists of a vast network of closed and open cavities in the form of membrane bound tubules, vesicles and flattened sacs and these are known as Endoplasmic reticulum or ER.
- The membranes are supposed to be originated from the inpushings of the plasma membrane.
- The membrane of ER attaches one side to the nuclear membrane and the other side to the plasma membrane and the internal space is called ER Lumen or ER cisternal space.

- The membrane of ER on its external surface carries granular structures known as ribosomes. The membrane of ER with ribosomes attached to its external side is called as Rough Endoplasmic reticulum or granular ER.
- The membrane of ER in certain regions may be devoid of ribosomes and they are called as smooth endoplasmic reticulum or agranular ER.

Rough ER Particularly developed in cells that are actively engaged in protein synthesis.
Present in pancreatic and liver cells

Function Synthesis of secretory proteins and are translocated through cisternae to different sites in the cells.

Smooth ER Found in regions rich in glycogen

Function Used in the formation of transport vesicles, which carry proteins and lipids to the Golgi complex.

*7. **Golgi complex (GC)***

- In the cytoplasmic matrix, a stack of flattened membrane bounded, parallely arranged organelles occur in the association of ER and is known as Golgi complex or Golgi apparatus (Fig. 2.6).
- Each GC is composed of many lamellae, tubules, vesicles and vacuoles
- The membranes of GC are made up of lipoproteins and these are supposed to be originated from the membranes of ER
- In plant cells, the GC are called as Dictyosomes and these secrete necessary material for cell wall formation during cell division

Function Storage of proteins and enzymes that are secreted by the ribosomes and transported by ER GC has the secretory function and it secretes lysosomes and many granules.

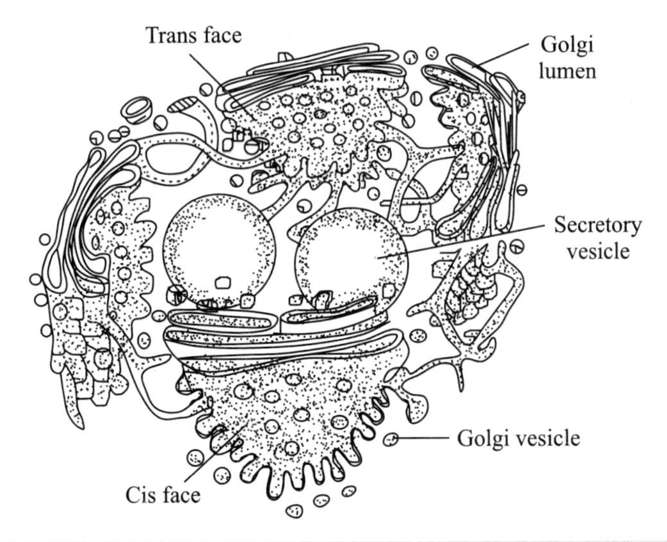

Fig. 2.6 Three Dimensional View of Golgi Apparatus (After Alberts etal., 1989)

8. Lysosomes The cytoplasm of the animal cells contain tiny spheroid, irregular shaped, membrane bound vesicles called lysosomes. Theses contain digestive enzymes that are capable of lysing and therefore it is called as lysosomes (a lytic body).

Five types of lysosomes were recognized according to the functions they perform and they are (i) Primary lysosome, (ii) secondary lysosome, (iii) Residual bodies, (iv) Autophagic vacuole and (v) Plant and fungal vacuoles (Fig. 2.7).

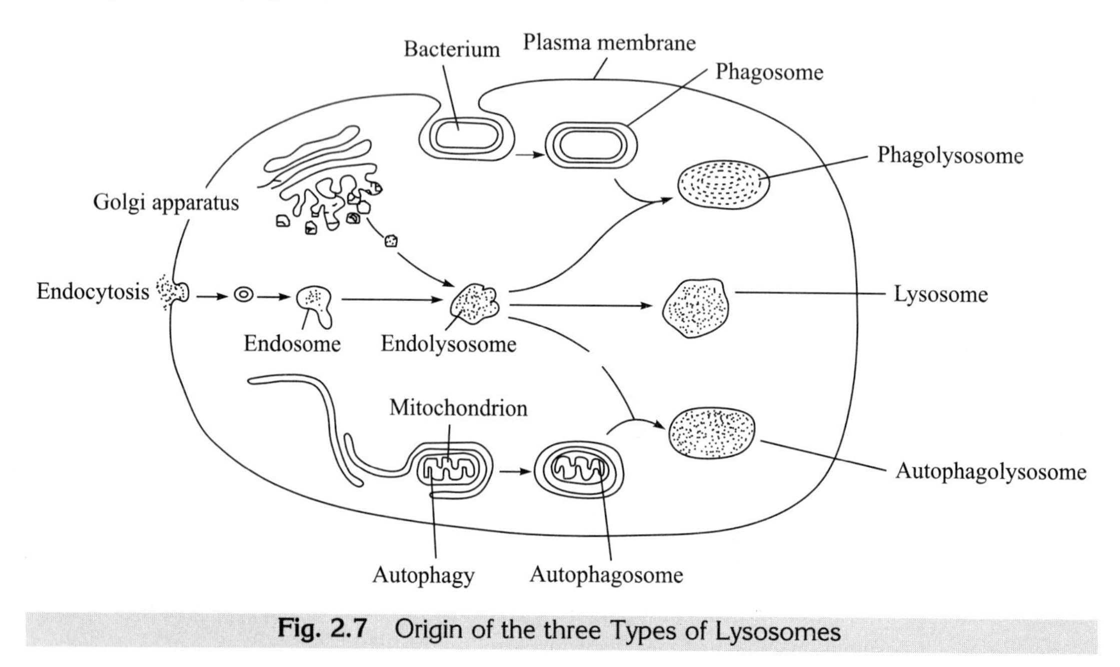

Fig. 2.7 Origin of the three Types of Lysosomes

Primary lysosomes are known as starch granule and is a small body having enzymes synthesized by ribosomes attached to ER. These enzymes first enter GC where an acid phosphatase reaction takes place.

Secondary lysosomes are heterophagosomes or digestive vacuoles resulting from phagocytosis or pinocytosis of foreign material by the cell. Hydrolytic enzymes digest the engulfed material.

Residual bodies are nothing but bodies that are formed when the digestion is incomplete. These are present in most of the protozoans and in other cases they persist and play a role in the ageing process. These are due to the absence of some enzymes and accumulation of phospholipids and these leads to severe pathologic disturbances.

Autophagic vacuole or autophagosome is a lysosome containing part of the cell itself for digestion. Plant and fungal vacuoles contain 30 – 90 per cent of the cell's volume compared to lysosomes in an animal cell. They contain hydrolytic enzymes but perform different functions such as:

- Storage of nutrients and waste products
- Degradation of waste material
- Increase of cell size
- Control of turgor pressure.

Functions

a. **Endocytosis and digestion of macromolecules**
   Hydrolysis starts in the endosomes having a pH of 6 and is completed in endosomes having a pH of 5. Common to all cells.
b. **Phagocytosis or digestion of external particles**
   Large particles and microorganisms are taken into the cell by a process called phagocytosis. The cell engulfs the particles and then forms and invagination that becomes pinched off from the cell membrane and forms an internal sac known as phagosome.
c. **Autophagy or digestion of intracellular substances**
   Digestion of the substances from the same cell to which the lysosome belong. Ex. degeneration of the tadpoles tail (enzyme called catepsin present in the lysosomes)
d. **Cellular digestion**
   When a cell dies, the lysosomal membrane gets ruptured and the enzymes are liberated. These enzymes digest the dead cell.
e. **Extracellular digestion**
   Reverse process of phagocytosis ex. Sperm penetrates the protective coating of the ovum.

9. Cytoplasmic vacuoles

- Hollow, large liquid filled structures
- Vacuole of plant cell are bounded by a single, semi-permeable membrane know as tonoplast
- Vacuole of animal cell are bounded by a lipo-proteinaceous membrane

Function Storage of products such as sugar and proteins

10. Microbodies The cytoplasmic matrix of many kinds of cells such as yeast, protozoan etc contain certain spherical, membrane-bounded particles measuring 0.3μm – 0.5μm diameter.

Function Hydrogen peroxide metabolism

11. Ribosomes

- Ribosomes are minute spherical structures attached to the surface of the ER forming rough or granular ER.
- Ribosomes originate from the nucleolus and consist of mainly RNA and proteins.
- Each ribosomes consists of two structural units namely the smaller one is the 40S Subunit and larger one is the 60S Subunit
- The ribosomes attach themselves to the ER by larger 60S subunit
- The 40S subunit occur on the larger subunit and form a caplike appearance
- Ribosomes consist of three types of RNA's known as ribosomal RNA's or 5S, 18S and 28SrRNA.
- The 28S and 5S occur on the large and 18S occur on the small sub-unit.

Function Serves as a major site of protein synthesis

12. Mitochondria

- In the cytoplasm of most cells occur large sized round or rod like structures called mitochondria.

- Mitochondria are bounded by two membranes made of lipoproteins
- The outer membrane forms a bag like structure around the inner membrane.
- The inner membrane gives out many finger like folds in the lumen of mitochondria and these folds of the inner mitochondrial membrane are called as Cristae
- The space between the outer and inner mitochondrial membrane is filled with a viscous mitochondrial matrix.
- The outer, inner membranes and the matrix are found to contain many oxidative enzymes and co-enzymes (Fig. 2.8).

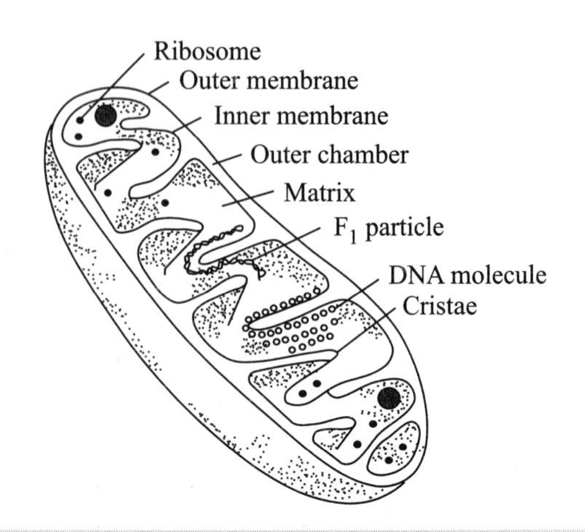

Fig. 2.8 Mitochondria – Longitudinal Section to Show the Internal Structure

Functions

- Respiration
- Oxidation of food
- Metabolism of energy
- Oxygen consumption in the cell
- Contain circular DNA molecule and ribosomes and are capable of synthesizing certain proteins (Fig. 2.9)

13. Plastids

- Occur commonly in plant cells
- Colorless or colored

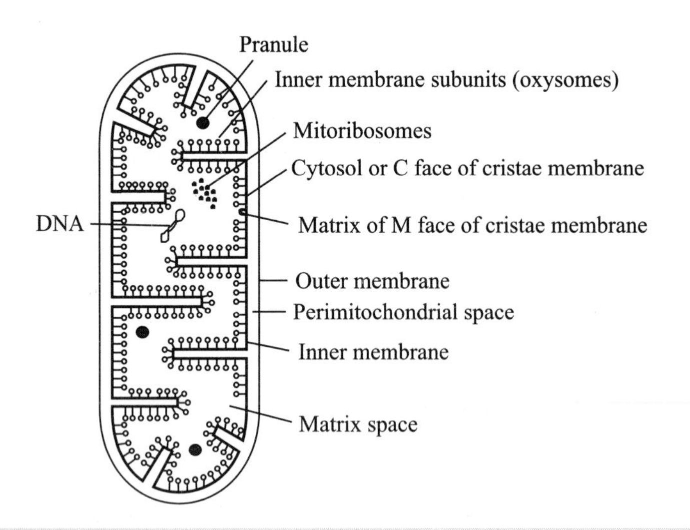

Fig. 2.9 Mitochondria – Sectional View of Oxysomes

- Colorless plastids are known as leukoplastids or leucoplasts
- Colored plastids are known as chromoplasts or chromo plastids

Function Storage of starch and lipids

14. Chloroplasts

- The chromoplasts having certain pigments particularly chlorophyll are known as chloroplasts (Fig. 2.10).
- Chloroplasts contain DNA, ribosomes and complete protein synthetic machinery.

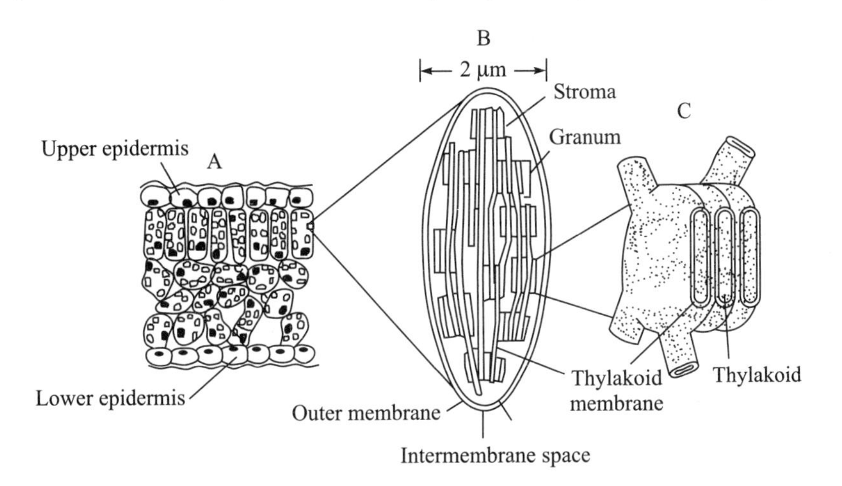

Fig. 2.10 A – Distribution of Chloroplasts in a leaf cell; B – Ultrastructure of a Chloroplast; C – Granum Details (After Alberts *et al.*, 1989)

Function Helps in the biosynthesis of food stuffs by the process of photosynthesis

15. Endosomes

- Endosomes are organelles made of heterogenous structure consisting of membrane bound tubules and vesicles
- Originate from endocytosis
- Have an acidic medium with pH ranging from 5 – 5.5.

Function

- Involved in intracellular traffic
- Helps in sorting of ligands and receptors

III. Nucleus

- Centrally located
- Spherical cellular component
- Carries the hereditary material – DNA
- Consists of three structures, viz

(i) Nuclear Membrane
(ii) Nucleoplasm and Chromosomes
(iii) Nucleolus

(i) Nuclear membrane

- Nucleus is bounded by two membranes made up of lipoprotein known as nuclear membrane
- Nuclear membrane forms a kind of envelope around the nucleus
- Nuclear membrane contains many pores through which transportation (in and out) takes place

Function

- Mechanical support
- Gives stability

(ii) Nucleoplasm and Chromosomes

- The space in-between nuclear membrane and nucleolus is filled with a watery substance known as nucleoplasm or Karyolymph.
- Nucleoplasm contains dissolved phosphorus, ribose sugars, proteins, nucleotides and nucleic acids.
- Nucleoplasm contains thread like elongated structures called chromosomes
- Chromosomes appear only during cell division or during the formation of chromatin granules
- Chromatin granules consist of large molecules of DNA and nucleoproteins.

(iii) Nucleolus

- Nucleoplasm contains a conspicuous, darkly stained spherical body known as nucleolus
- Chemically, nucleoplasm consists of large amount of ribosomal proteins and ribosomal RNA.

Functions

- Stores the rRNA molecules
- Biogenesis of ribosomes

2.3 VIRUS – ITS UNIQUENESS

Viruses are infectious particles that consist of a DNA or an RNA molecule (the viral genome) packed in a protein capsid, which in the enveloped viruses is surrounded by a lipid bilayer – based membrane. Both the structure of the viral genome and its mode of replication vary widely among viruses. A virus can multiply only inside a host cell whose genetic mechanisms it subverts for its own replication. A common outcome of a viral infection is the lysis of the infected cell and release of infectious viral particle. Many viruses are thought to have evolved from plasmid, which are self-replicating DNA or RNA molecules that lack the ability to wrap themselves in a protein coat.

Viruses are parasitic living organisms placed between prokaryotic and eukaryotic cells and also between living and non-living things. When viruses live inside the cell, they are active and they feed, respire, reproduce, grow and move. When they live outside, they remain inactive and do not feed, respire, reproduce, grow and move. Thus, the viruses resemble the living organisms in the intra-cellular space and the non-living organisms (chemicals) in the extra-cellular space.

Table 1 Comparison of Prokaryotic and Eukaryotic Organisms

	Prokaryotes	*Eukarotes*
Organisms	*Bacteria and cyanobacteria*	*Protests, fungi, plants and animals*
Cell Size	Generally 1 to 10 μm in linear dimension	Generally 5 to 100 μm in linear dimension
Metabolism	Anaerobic or aerobic	Aerobic
Organelles	Few or none	Nucleus, mitochondria, chloroplasts, endoplasmic reticulum etc.
DNA	Circular DNA in cytoplasm	Very long linear DNA molecules containing many non-coding regions bounded by nuclear envelope
RNA & Protein	RNA and protein synthesized in the same compartment	RNA synthesized and processed in nucleus and proteins synthesized in cytoplasm
Cytoplasm	No cytoskeleton. Cytoplasmic streaming, endocytosis and exocytosis - absent	Cytoskeleton composed of protein filaments. Cytoplasmic streaming, endocytosis and exocytosis
Cell Division	Chromosomes pulled apart by attachments to plasma membrane	Chromosomes pulled apart by cytoskeletal spindle apparatus
Cellular organization	Mainly unicellular	Mainly multicellular with differentiation of many cell types

Viruses are Mobile Genetic Elements

Viruses were first thought to be disease-causing agents that can multiply only within the cells and later found that it can also move from one cell to another cell. The idea that viruses and genes carry our similar functions was proved by studies on bacteriophages. Bacteriophages are bacterial viruses and later it was studied that the bacteriophage, T4 Phage DNA and not the protein enters the bacterial host cell and initiates the replication events that leads to the production of several hundred progeny viruses in every infected cell. These observations led to the idea that viruses are genetic elements. Moreover, the nucleic acid in the virus, the structure of it's coat, it's mode of entry into the host cell and its mechanism of replication inside the cell, all vary from one type of virus to another.

Structure of a Virus

Virus is made up of a central core of **nucleic acid** covered with a sheath called **capsid**. The nucleic acid may be either **DNA** or **RNA**. When the nucleic acid is DNA in a virus, it is called **DNA-Virus** and when the nucleic acid is RNA in a virus, it is called **RNA-Virus**. Therefore, the nucleic acid in the virus may be either **single-stranded** or **double stranded** (genetic material).

The outer coat of a virus is a protein capsid or a membrane envelope

The capsid is formed of a protein, composed of numerous protein subunits called capsomers. Capsomeres have different shapes in different viruses. The function of the capsid is to protect the virus when it remains outside the cell and this capsid is absent in plant viruses.

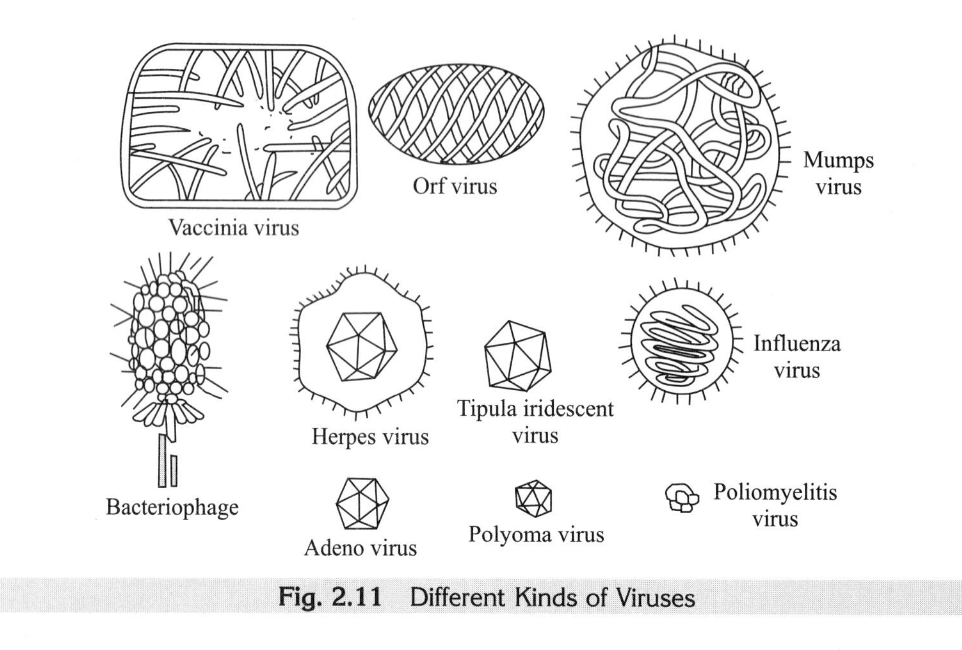

Fig. 2.11 Different Kinds of Viruses

Viral genomes can be either RNA or DNA

The DNA double helix has the advantages of stability and easy repair. If one polypeptide chain is accidentally damaged, its complementary chain permits the damage to be readily corrected. Therefore, the genetic information in a virus can be carried out in a variety of unusual forms, including RNA instead of DNA. A viral chromosome can be a single-stranded RNA chain, a double stranded RNA helix, a circular single stranded DNA chain, or a linear single stranded DNA chain.

A viral chromosome codes for enzymes involved in the replication of its nucleic acid

Each viral genome requires unique enzymes for replication and must encode not only the viral coat protein but also one or more of the enzymes needed to replicate the viral nucleic acid.

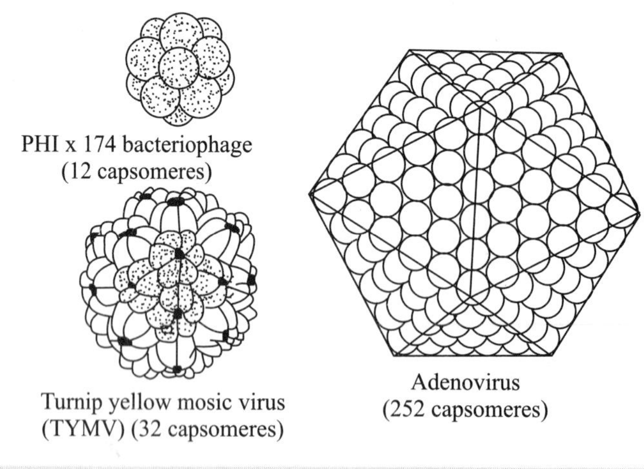

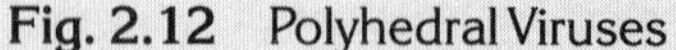

Fig. 2.12 Polyhedral Viruses

DNA and RNA viruses replicate by forming complementary strands

This formation of complementary strands is a process catalyzed by specific RNA dependent RNA polymerase enzymes.

Classification

Viruses are generally classified into three types namely

A. Plant viruses
B. Animal Viruses
C. Bacteriophages

A. Plant viruses

Plant viruses have RNA as the genetic material. Example – TMV (Tobacco Mosaic Virus) (Fig. 2.13)

B. Animal viruses

Animal viruses have DNA as the genetic material. Example – Bacteriophage (small pox virus, influenza, mumps etc)

C. Bacteriophages

Bacteriophages attach themselves to bacteria and are also called as bacterial virus. Example – E.coli (or) T_4 Bacteriophage

Entry of Virus into a Cell – Example: Semilike Forest Virus

All viruses have a limited amount of nucleic acid in their genome and so they parasitize host-cell pathways for most of the steps in their reproduction. Because viral products are usually synthesized in large amounts during infection, and because during its life cycle the virus follows a sequential route through the compartments of the host cell, virus-infected cells have served as important models for tracing the

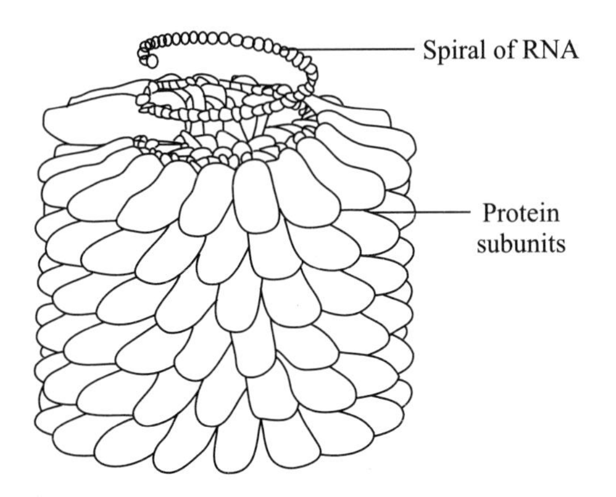

Fig. 2.13 Molecular Organization of the Tobacco Mosaic Virus (TMV) (After DeRobertis and De Robertis, 1987)

pathways of intracellular transport and for studying how essential biosynthetic reactions are compartmentalized in eukaryotic cells.

Enveloped animal viruses, in which the genome is enclosed in a lipid-bilayer membrane, have exploited the compartmentalization of the cell to an especially fine degree. To follow the cycle of an enveloped virus, we shall study the Semilike forest virus as an example.

The **Semiliki forest virus** is a single stranded RNA genome surrounded by a capsid formed by a regularly arranged icosahedral (20 faced) shell composed of many copies of one protein called C protein. The nucleocapsid (genome + capsid) is surrounded by a closely opposed lipid bilayer that contains only three types of polypeptide chains, each encoded by the viral RNA. The envelope proteins form heterotrimers that span the lipid bilayer and interact with the C protein of the nucleocapsid linking the membrane and nucleocapsid together. The glycosylated portions of the envelope proteins are always on the outside of the lipid bilayer and each trimer forms a "spike" that can be viewed through electron micrographs projecting outward from the surface of the virus.

Infection is initiated when an envelope protein on the virus binds to a normal cell protein that serves as its receptor on the host-cell plasma membrane. The virus than uses the cell's normal endocytic pathway to enter the cell by receptor-mediated endocytosis and is delivered to early endosomes. Then, instead of being transferred from endosomes to lysosomes, the virus escapes from the endosome by virtue of the special properties of one of its envelope proteins. At the acidic pH of the endosome, this protein causes the viral envelope to fuse with the endosome membrane, releasing the bare nucleocapsid into the cytosol. The nucleocapsid is "uncoated" in the cytosol releasing the viral RNA, which is then translated by host-

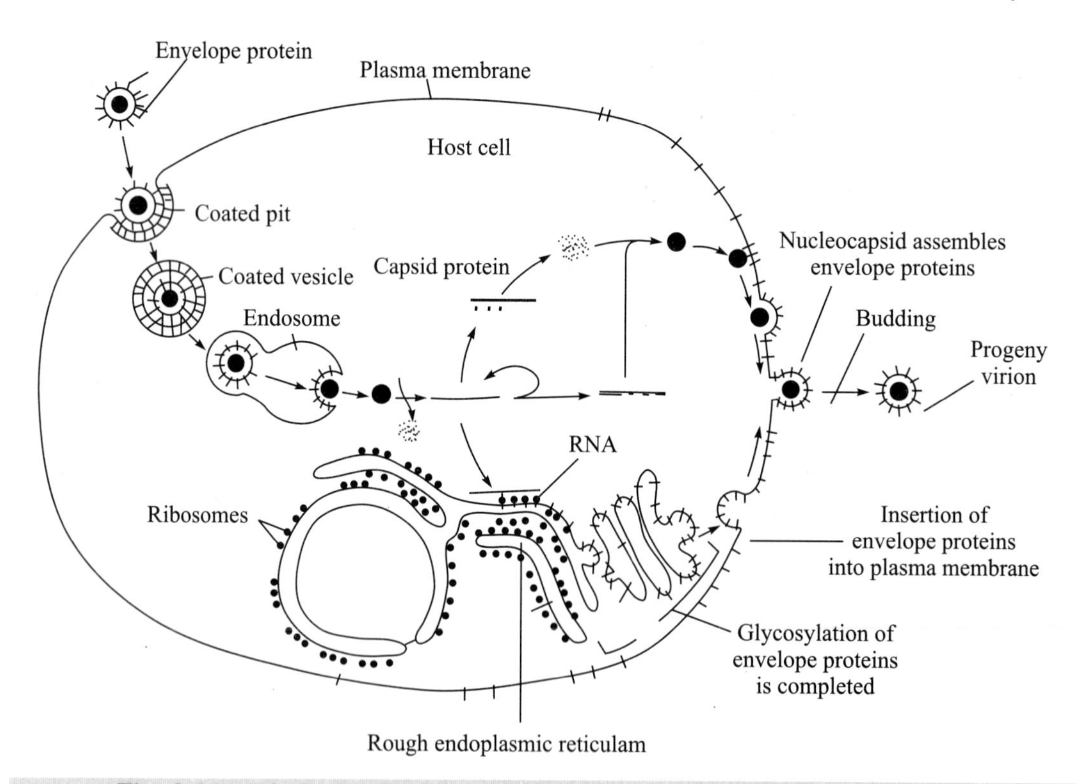

Fig. 2.14 Life Cycle of the Semiliki Forest Virus (After Alberts *et al.*, 1989)

cell ribosomes to produce a virus –encoded RNA polymerase. This in turn makes many copies of viral RNA, some of which serve as mRNA molecules to direct the synthesis of the structural proteins of the virus – the capsid C protein and the three envelope proteins.

The newly synthesized capsid and envelope proteins follow separate pathways through the cytoplasm. The envelope proteins, like the plasma membrane proteins, of the host cell are synthesized by ribosomes that are bound to the rough ER. In contrast, the capsid protein, like the cytosolic proteins of the cell is synthesized by ribosomes that are not membrane bound. The newly synthesized capsid proteins bind to the recently replicated viral RNA to form new nucleocapsids. The envelope proteins, in contrast, are inserted into the membrane of endoplasmic reticulum, where they are glycosylated and transported into the Golgi apparatus. From the Golgi apparatus, they are finally delivered into the plasma membrane.

The viral nucleocapsids and envelope proteins finally meet at the plasma membrane. As a result of a specific interaction with a cluster of envelope proteins, the nucleocapsid forms a bud whose envelope contains the envelope proteins embedded in host-cell lipids. Finally, the bud pinches off and a free virus is released on the outside of the cell. The clustering of envelope proteins as they assemble around the nucleocapsid during viral budding excludes the host plasma membrane proteins from the final virus particle.

2.4 FUNCTIONS OF CELL

The functions of the plasma membrane are grouped as

1. Endocytosis and Exocytosis
2. Chemotaxis
3. Cell Movement
4. Cell adhesion

Endocytosis and Exocytosis

Endocytosis

The material to be transported from the outside of the cell towards the interior of the cell, occurs by a process of forming an invagination and progressively pinches off into a vesicle towards the lysosome and are called **endocytosis.** In the process of endocytosis, the engulfment of the solid / liquid material after getting pinched off into vesicles are called as **endosomes**. When the engulfed particles are large, generally (greater than 250 nm. in diameter), they are called as **phagosomes**.

Endocytosis is classified into **two types** namely

A. Phagocytosis
B. Pinocytosis

A. Phagocytosis Phagocytosis is a special form of endocytosis, through which the ingestion of large particles through phagosome takes place by specialized phagocytic cells and this process is also called as 'cell eating'.

Example

Amoeba A pocket like cavity, lined with clathrin, is formed and the ingestion of prey occurs. The plasma membrane closes after ingestion forming a vesicle and that travels to the lysosome for other dehydration functions or is reverted back to the plasma membrane.

B. Pinocytosis Pinocytosis is a special form of endocytosis and it is the intake of **fluid materials**, which are unable to pass through the membrane, and this process is also called as 'cell drinking'.

Example

Amoeba The plasma membrane folds inside forming a pocket and the fluid droplet passes into it. The pocket deepens and finally nipps off as a fluid filled vacuole (or) **pinocytic vacuole** called as **pinosome.** The pinosome passes to the interior of the cell and here, either it fuses with the lysosome and the materials are digested and the digested materials diffuse into the cytoplasm. The cell takes inside the dissolved materials such as ions, sugars, fatty acids and amino acids by pinocytosis.

Exocytosis

Exocytosis is the reverse process of endocytosis and this is a process by which the cell throws out its products and this process is otherwise called as **'cell vomiting'** (or) **'reverse endocytosis'**.

Example

Pancreatic gland cells and sweats gland cells The secretory products move from the interior to the surface of the membrane and here, they fuse with the plasma membrane and discharge the contents outside the cell. Undigested food materials throw out the excess food through the form of exocytosis in lower animals called amoeba and paramecium.

Chemotaxis

Chemotaxis is defined as movement in a direction controlled by a gradient of a diffusible chemical sensed by the cell. One well-studied example is the chemotactic movement of certain white blood cells (neutrophils) toward a source of bacterial infection. Neutrophils have receptor proteins on their surface that enables them to detect the very low concentration of the N-formylated peptides derived from bacterial proteins.

Another well-studied example of chemotaxis is provided by the cellular slime mold, *Dictyostelium discoideum.* These eukaryotes live on the forest floor as independent motile cells called amoebae, which feed on bacteria and yeast and under optimal conditions, divide every few hours. When their food supply is exhausted, the amoebae stop dividing and gather together to form tiny, multicellular, worm like structures, which crawl about as slugs and leave trails of slime behind them. As the slug migrates, the cells begin to differentiate, initiating a process that ends with a production of a tiny plantlike structure consisting of a stalk and a fruiting body some thirty hours after the beginning of aggregation. The fruiting body contains large number of spores, which can survive for long periods of time even in extremely hostile environments. Only when conditions are favorable, the spores germinate to produce the free-living amoebae that start the life cycle again. The Dictyostelium amoebae aggregate by chemotaxis, migrating toward a source of cyclic AMP, which is secreted by the starved amoebae. Like neutrophils, the amoebae reorient their leading edge in order to migrate up a shallow chemo attractant gradient. When they are exposed to a local cyclic AMP, leaking from a micropipette, they extend actin-containing processes directly toward the pipette.

Cell Movement

All living organisms have to perform the vital function of transportation and all cell and cell organelles allow transport of substances into and outside these structures.

Table 2 Different Types of Molecules and their Possible Transport through the Plasma Membrane

S.No.	*Types of molecules*	*Examples*	*Transport Mechanism*
1	Hydrophobic molar molecules (non-polar)	O_2, CO_2, N_2	Rapid diffusion
2	Uncharged small molar molecules (Polar)	H_2O, Urea, Glycerol	Rapid diffusion
3	Uncharged large polar molecules	Glucose, Sucrose	Hardly diffuse. Move through pores using carrier proteins
4	Charged Molecules (Polar)	H^+, Na^+, K^+, Ca^{2+}, $Mg2^{2+}$, Cl^-	No diffusion. Move using ion channels (or) active transport

Uncharged large polar molecules and charged polar molecules do not undergo diffusion and they move across membrane using the regulated pores / channels / ions / and by transport proteins or pumps found in the membrane itself. Cells can transfer large (or) macromolecules across the membrane through processes such as exocytosis and endocytosis. These are done by the formation of vesicles and these are otherwise called as vesicular transport.

Membrane proteins (or) transport proteins are of two type, namely,

(1) Carrier proteins and
(2) Channel proteins.

Membrane transport is of two types, namely,

(1) Passive transport (includes passive diffusion and facilitated diffusion)
(2) Active transport

Membrane excitability is defined as follows: -

Membrane transport is the response of cells to external and internal stimuli, which involves muscle movement and rapid inter-cellular communications. These communications make use of receptor proteins in the membranes of excitable cells. Ex. nerve cells or neurons or neurotransmitters.

Cell Adhesion

The cells in the extra cellular matrix are held intact by **cell-cell adhesion** resulting in the formation of structures called as **cell junctions**. Cell sorting is a phenomenon that allows different types of cells to segregate and form specific regions of a developing embryo.

Townes and Holfreter provided a definite example for cell sorting in the year 1955. They demonstrated that 'Dissociated cells of any vertebrate embryonic organs can re-aggregate and sort out to re-construct semblances of the original structure. **Example – Amphibian tissues** (become dissociated into single cells when subjected to an alkaline medium. Suspended cells are found with prominent germ layers such as ectoderm, mesoderm and endoderm). When cells of different types are mixed, cells of the same type aggregate together through the phenomenon of cell adhesion. The results were interpreted in terms of **selective affinity**.

Selective affinity in cells of different types aggregate as ectoderm having +ve affinity with mesoderm and –ve affinity with endoderm and mesoderm having +ve affinity with both endo and ectoderms.

Example

During onset of gastrulation specific group of cells at the vegetal end of the blastula lose affinity with the neighboring cells and ECM and acquire affinity for protein fibrils where they form into skeletal structures i.e. having significant role in morphogenesis.

Cell adhesion molecules (CAMs)

Different cells interact among themselves as well as to the external environment during differentiation and morphogenesis of an organism. This is done through Cell Recognition and Cell Adhesion and thus mediated by

1. Cell adhesion molecules
2. Cell substratum molecules and
3. Cell junction molecules.

Cell adhesion molecules include 3 major receptors:

- Cadherins – adhesion properties are dependent on Ca^{++} ions
- Immunoglobulin super family (Ig) super family - Ca^{++} independent adhesion; includes neural cell – cell adhesion moles (N-CAM) and intercellular adhesion moles (I-CAM)
- Selectins

Cell substratum adhesion molecules include 1 major receptor:

- Integrins

(I) Cadherins These are Ca-dependent glycoproteins – crucial for the spatial segregation of cells leading to the organization of an organism. Ex. In mammalian embryos, 3 classical cadherins are found. They are

i. E-cadherins – epithelial cadherins (or uvomorulins or L-CAM),
ii. P-cadherins – placental cadherins, and
iii. N-cadherins – neural cadherins (or A-CAM)

Specific cells adhering to other cells expressing the same cadherins exhibit a homophilic **binding. Heterophilic binding** through an extracellular linker molecule is also possible.

Each cadherin has 3 major regions and they are the

a. Extracellular domain of 113 amino acids - an NH_2 terminal that helps in cell sorting or cell recognition and are responsible for Calcium binding
b. A Transmembrane domain, and
c. Cytoplasmic domain that mediates interaction with at least 3 cytosolic proteins such as alpha, beta and gamma **catetins**.

In the presence of Ca ions, the cytoplasmic domain of a cadherin is connected to actin filaments through catenins. The clustering of cadherins, catenins and actins leads to the formation of a belt-like **adherens junction**, which is responsible for cell – cell adhesion. The integrity of the junction is regulated by protein-tyrosine phosphorylation. Cell – cell adhesion is also represented by desmosomes – mediated by 3 transmembrane Ca dependent proteins such as **Desmoglein I, Desmocollin I & Desmocollin II.**

In case of cytoplasmic domains, it is mediated through a cytoplasmic structure called **desmosomal plaque**.

Cadherins are highly important in establishment and maintenance of cell-cell interactions. The deficiency of cadherins may lead to malignancy of cells – life threatening diseases such as Pemphigus **vulgaris** where epidermal cells lose adhesion, leading to blister formation and skin fall off.

(II) Immunoglobulin Superfamily (Ig) Ca++ independent cell adhesion molecules, most populous and represent a functionally diverse family of molecules present on the cell surface. Ig related CAMs were first discovered in a unicellular Dictyostelium (slime mold). When this slime mold is allowed to starve, it aggregates into a multicellular organism by the help of CAMs.

80kD Glycoprotein gene made its expression in vegetatively dividing cells when antigen-binding sites were isolated from antibodies. During division, the cells started adhering to each other although in vegetative phase, they should not adhere. This led to the discovery of an 80kD glycoprotein that mediates cell-cell adhesion during a slime mold aggregation.

140kD N-CAM (Neural Cell Adhesion Molecule) was isolated from Chick embryos using the same technique of isolating antigen-binding sites. This 140kD N-CAM plays an important role in neuronal aggregation.

N-CAM is the best-studied example of Ig Super family CAMs and this in

- Found in a variety of cell types such as nerve cells that bind adjoining cells together by homophilic binding.
- Other Ig like CAMs that use heterophilic binding (ICAM) – Intercellular Adhesion Molecules found on endothelial cells. These ICAMs bind to integrins (occur on the surface of WBC) to help trapping of white blood cells at the time of inflammation.
- There are at least 20 forms of N-CAMs.
- Each N-CAM consist of as an

  1. Extra cellular domain folded into five Ig-like domains,
  2. transmembrane Domain spanning the membrane and an
  3. Intracellular or cytoplasmic domain.

Some N-CAMs have extra cellular domain only and others are attached to the plasma membrane by a glycosylphosphatidyl-inositol (GPI) molecule and others are found to occur in all the three domains, varying in sizes and differing in the degree of glycosylation. There are Ig like proteins that mediate Ca++ independent cell-cell adhesion has been studied in Drosophila. Ex. Fasciclin II.

Fasciclin II is a close relative of N-CAM and is expressed on a subset of nerve cell processes and on some of glial cells.

N-CAMs differ in the quantity of associated sialic acid. N-CAMs and N-Cadherins are found on the same cells but in different region. The adhesiveness of the cells with N-CAM molecule varies with the amount of sialic acid. Cells with low amount of sialic acid on their N-CAMs aggregate four times more readily than those with a high level of sialic acid. As the embryo gets older, most of the N-CAM proteins progress from high sialic acid to low sialic acid forms which stabilizes matured tissues. If two neighboring cells express low sialic acid form of N-CAMs – strong cell adhesion is promoted whereas in cells containing high sialic acid N-CAMs, cell adhesion is weakened or inhibited.

Segregation of different kinds of cells is also achieved by the presence of different CAMs on these different kinds of cells.

For example,

- Notochord cells do not enter into a neural tube
- Dermal cells do not enter epidermis layer
- Development of feathers in a chicken
- Connections between an axon (axons are extensions of nerve cell body) – fasciculation

(III) Selectins

- Selectins are a family of three cell surface glycoproteins that contain lectin domains for recognizing cell-specific glycoproteins during regional inflammation.
- Selectins provide a mechanism for cell-cell adhesion, where lectin or carbohydrate recognition domain (CRD) in a selectin of one cell recognizes a cell surface carbohydrate of another cell.
- Unlike other CAMs, selectins facilitate adhesion and encourage segregation of cell types.
- Three known selectins are there namely
  (a) L-Selectin or Leukocyte-Selectin,
  (b) E-Selectin or Endothelial –Selectin, and
  (c) P-Selectin or Platelet-Selectin
- Detailed studies of the above three selectins suggested that excellent correspondence between the domain structure. The domains in these selectins include:
  (a) A Lectin domain (L) or Carbohydrate recognition domain,
  (b) An CRD epidermal growth factor (EGF) like Motiff (E), and
  (c) A variable number of repetitive domains of glycoprotein C.
- Similarly, in an overall way exists between L and E domains (high)
- Less similarity in C domains

Table 3 Characteristic Features

S.No.	*Type*	*Location*	*Expression*	*Function*
1.	L-Selectin	Leucocytes	Decreases on cell activation	Lymphocyte recirculation through PLN. Inflammation
2.	E-Selectin	Endothelium	Increases on inflammation	Leucocyte inflammation
3.	P-Selectin	Platelets	Increases upon thrombin activation	Leucocyte inflammation

(IV) Integrins

Terminology Named by Horwitz et al., 1986 & Tamkun et al., 1986

Definition Integrins are transmembrane adhesion-receptor proteins

Function

- They integrate extra and intracellular contents
- They mediate cell-substratum adhesion done by fibronectin and **vitronectin** (Extracellular) and by **actin filament** bundles (intracellular)

Occurrence Integrins are found in the focal adhesion areas of the plasma membrane closely with the substratum (10-15nm gap).

Structure & Characteristics

- Integrins have alpha (14 subunits of 120-180kD) and beta (8 subunits of 90-110kD) heterodimers (20 subunits) that are non-covalently linked.
- Heterodimers facilitate binding of extracellular moles, allowing cell-cell and cell-ECM interaction
- Integrins have many sub-families based on beta chain component
- The largest subfamily is β (beta)1 or VLA integrin (very late activation antigen)
- There are 7 different alpha subunits, sharing common beta chain
- Four of the β (beta)1 integrins bind to **laminin, collagen** and **laminin** constituting ECM
- β (Beta) 2 consists of leucocyte receptors, that involve in leucocyte-leucocyte or endothelial cell or monocyte-endothelial interaction
- β (Beta) 3 consists of platelet glycoprotein, fibrinogen receptor and vitronectin receptor
- Vitronectin is multifunctional glycoprotein in circulation tissue, amniotic fluid and urine; helps cell adhesion, defense and cell invasion
- Though alpha and beta integrins facilitate focal adhesion only the beta subunit connects internal cytoskeleton
- Other cytoplasmic domain proteins involved in adhesion are – talin, alpha-actinin, vinculin, paxillin, zyxin, actin, redixin, ezrin, moesin, tenuin and VASP (vasodilator stimulated phosphoprotein).
- Integrins also play a major role in development through tissue and cell adhesion'.
- Ex. 2 integrins PS1 and PS2 help in the development of *Drosophila melanogaster* wing
- Usually the tissues are attached to each other and the membrane; but the circulating cells are without attachment – This attachment and detachment are essential for proper growth, development and survival of cells, as well
- But without adhesion cells undergo programmed cell death – apoptosis. Therefore, cell attachment is regarded as anchorage dependence and apoptosis as anoikis
- Focal adhesion is lesser when the rate of cell movement on the substratum is more (inversely proportional)
- Integrins utilize focal adhesion kinase enzyme for control of anchorage – therefore integrins activate many signalling molecules
- Anchorage dependence is also regulated by cell shape – flat cells thrive better than rounded cells
- Cells forced to extend over a larger surface area survive better and proliferate faster
- ECM controls cell shape whereas cell shape controls survival and growth of cell
- This involves activation and inactivation of certain pathways – integrins facilitate signals for survival in flat cells but apotosis in rounded cells
- In a damaged tissue, fewer cells cover the area spreading to larger surface stimulating proliferation till the gap is filled

Glycosyltransferase

- Glycosyltransferases are found in endoplasmic reticulum and Golgi vesicles

- Glycosyltransferases are meant for adding different sugars to peptides forming glycoproteins
- *Ex. Laminins that lead to adhesion*
- Glycosyltransferases also help in fertilization – Glycosyltransferases in sperm cell membrane interacts with the carbohydrate component of ECM secreted by the egg
- Glycosyltransferases are involved in migration of cells – sugar is transferred to the substrate on which the cells migrate

Cell Junctions

These are found in animal tissues among cells and between cells and ECM – to facilitate cell-cell and cell-matrix contacts (in plants called as 'plasmodesmata)

Abundant in epithelia – The cell-cell and cell-matrix junctions are collectively termed as cell junctions. There are 3 functional types:

1. **Occluding junctions (tight junctions)** – sealing cells together in an epithelial sheet preventing passage of molecules from one side to other
2. **Anchoring Junctions** – allowing attachment of cells to neighboring cells or ECM
3. **Communicating Junctions** – allowing chemical or electrical signals to pass from cell to cell

1. Occluding junctions

Function as a selective barrier between fluids. Ex. In mammalian small intestine the epithelial lining does not allow the contents to pass on the lumen but instead absorb the nutrients. After permeation, the nutrients get into the blood vessels. 2 carrier proteins do this transport and they are

1. Carrier proteins confined to the **apical surface** – from lumen to cells (sodium driven glucose carrier) and
2. Carrier proteins confined to the **basolateral surface** – allow passage of moles from cell to extra cellular fluid by facilitated diffusion (by glucose carrier)

The diffused moles are not allowed to diffuse back from intercellular spaces. The permeability of epithelia to small moles differs. The epithelia of small intestine are 10,000 times more permeable to inorganic ions than the epithelium lining urinary bladder.

The tight junctions consist of anatomizing network of strands with specific transmembrane proteins.

2. Anchoring junctions

- They help in connecting cytoskeletal elements of one cell to another (cell-cell adherens junction) or to ECM (cell-matrix adherens junctions)
- Most common in heart muscle or skin epithelium that is subject to mechanical stress.
- Anchoring Junctions include: -

  a. **Adherens** junctions: connection sites for actin filaments

  b. **Desmosomes and hemidesmosomes** – connection sites for intermediate filaments. They are composed of **intracellular attachment proteins** and **transmembrane linker proteins** like cadherins and integrins.

  c. **Septate** junctions: unique to invertebrates and function as connecting sites for actin filaments.

3. Communicating junctions

Communicating Junctions are meant for the passage of chemical or electrical substances (e.x. ions). These include gap junctions, chemical synapses and plasmodesmata. Gap junctions and chemical synapses are common in animals and plasmodesmata are common in plants only.

Gap junction in animals Cell – Cell attachments and communications are facilitated by proteins forming 'gap-junctions' between cells that maintain a gap of 2nm to 4nm wide. In tight junctions, on the other hand, plasma membranes of two cells appear to be in direct contact without any gap. These gap junctions are common in most of the animal tissues and cells. These can be observed under an electron microscope only as patches between cells separated by a gap.

The gap junction channels are made up of structures called 'connexons' which are made up of transmembrane proteins called 'connexins'. Six identical connexins group together to form a transmembrane channel containing a central pore. With a connexon comes in contact with a similar channel from an adjoining cell, a gap junction is formed allowing both the cytoplasm to communicate among themselves.

The gap junction has each pore of 1.5nm in diameter and through this pore; small water-soluble molecules (not bigger than 1000 Daltons) pass from cytoplasm of one cell to that of another cell. These cells connected by gap junctions are described as electrically coupled when ions are passed through gap junctions or metabolically coupled.

Cell junction in plants – plasmodesmata The plasmodesmata are fine cytoplasmic channels between cells and these bring cytoplasm of adjoining cells to come together. Running through the center of this channel is a narrow cylindrical structure called desmotubule, which is continuous with the endoplasmic reticulum of the adjoining cells. Between the inner wall of plasmodesmata and the desmotubule is the cytosolic annulus through which molecules can pass.

Plasmodesmata resemble gap junction in several ways but are contrast in having elongated structures that traverse the thick cell wall.

Each plasmodesmata has an

(1) outer sheath continuous with the plasma membrane a
(2) central core of endoplasmic reticulum and
(3) collar or neck region.

Function

- Perform passive diffusion of permitting small metabolites between plant cells
- Larger genomes of plant viruses can traverse from one cell to another
- They rapidly alter their structures to allow transport of larger molecules
- Endogenous plant protein transport
- Movement proteins of plant viruses have been discovered and this operates on endogenous plasmodesmata transport system. If Membrane Proteins are injected into the single cells, the permeability of Plasmodesmata is increased within minutes of microinjection.
- It has been confirmed that plasmodesmata in different cell types in plants are diverse and therefore, respond differently to membrane proteins in different cell types.
- Membrane protein has a transport signal, which dictates a transport irrespective of the size of the molecule to be transported. The four criteria for a molecule to be transported include size, shape, and signal sequence and gating junction.

Questions

1. Differentiate a Prokaryotic and Eukaryotic cell with an example:-
2. Draw and mark the parts of a typical Eukaryotic cell:-
3. What are plasma membranes ? Mention any two functions of it:-
4. What are the Cytoplasmic structures in the cell?
5. What are Microtubules, Cytoplasmic Filaments and Centrosomes? Give two functions of each:-
6. What are basal granules (or) Kinetosomes?
7. What is GC? What is its important function?
8. What are lysosomes? State its classification based on its function:-
9. What are Microbodies, Ribosomes, Chloroplasts and Mitochondria?
10. Define Nucleus, Nuclear membrane and Nucleoplasm:-
11. Explain virus as a mobile genetic element and outline its entry into a host:-
12. What is a capsid?
13. Classify viruses?
14. Define endocytosis and exocytosis:-
15. Define chemotaxis:-
16. Differentiate cell adhesion and cell movement:-
17. What are carrier proteins and channel proteins?
18. Explain passive transport and active transport:-
19. What is selective affinity?
20. What are the major receptors in cell adhesion?
21. Explain Cadherins, Immunoglobulin and Selectins?
22. Define Cell junction, Occluding junction and Communicating junction:-
23. What is a Plasmodesmata? Explain:-
24. What are membrane proteins?
25. What is a Semiliki Forest Virus?

CHAPTER 3

Molecular Biology of the Cell

3.1 MEMBRANE PROTEINS

Membrane Proteins are proteins associated with the plasma membrane either by its location, function or by any other related means. Proteins constitute 70 % of the plasma membrane and the major functions of the plasma membrane are attributed to the proteins.

Membrane Proteins **exist in a variety of forms** that include

1. Transmembrane Proteins
2. Covalently linked Cytosolic Extrinsic Proteins
3. Covalently linked Non-cytosolic Extrinsic Proteins
4. Non-Covalently linked Extrinsic Proteins
5. Integral or Intrinsic Proteins
6. Representative Membrane Proteins
    - 6.1 Spectrin
    - 6.2 Glycophorin
    - 6.3 Band 3
7. Bacteriorhodopsin
8. Porins
9. Protein Complex in membranes
10. Specific Proteins for Specific Membrane Domains
    - 10.1 Cell coat
    - 10.2 CAMs (Cell Adhesion Molecules)
        - 10.2.1 Cadherins
        - 10.2.2 Integrins
        - 10.2.3 Selectins
        - 10.2.4 N-CAM

1. Transmembrane Proteins

- Extend through the lipid bilayer as a single helix. Example – Glycophorin
- Extend through the lipid bilayer as multiple helices (or) as barrels.

Example - Porin

- When the polypeptide crosses the membrane only once, it is called as 'single pass transmembrane'. Example – Glycophorin protein of human RBC's.
- When the polypeptide crosses several times, it is called 'multipass transmembrane protein'. Example – Band 3 protein of human RBC's.
- All Transmembrane proteins pass the entire bilayer and there is no certain established example of a protein, which is embedded only across the lipid bilayer
- The transmembrane regions of these proteins are hydrophobic and they interact with hydrophobic tails of lipid molecules.
- One or both the regions of these proteins lying outside the membrane are hydrophilic, exposed to water.

Hydropathy plot

Hydropathy plot is a graph used to estimate the number of α-helical peaks and hydrophobicity across the number of amino acids.

Example – Glycophorin – Single peak (Single α-helix)
Bacteriorhodopsin – Multiple Peaks (7 α-helices)

Transmembrane proteins are difficult to crystallize and cannot be subjected to x-ray crystallography to study its three dimensional structure and properties. However, cloning and sequencing of genes encoding the properties give valuable information about the nature of α and β strand traversing the lipid bilayer. The segments containing 20 – 30 amino acids with a high degree of hydrophobicity are long in the α -chain and for 10 amino acids in the β chain.

2. Covalently Linked Cytosolic Extrinsic Proteins

- The hydrophobicity of membrane proteins is increased by covalent attachment of a fatty acid chain that is inserted in the cytoplasmic leaflet of the lipid bilayer.
- Located entirely in the cytosol
- Attached to the membrane by means of covalently attached fatty acid chains or phenyl groups.
- Proteins + amino group – Amide linkage
- Proteins + Thioether linkages –Phenyl group
- Fatty acid yields an amide linkage between terminal amino group and fatty acid
- Phenyl group yields a thioether linkage between cysteine and Phenyl group.

3. Covalently Linked Non-cytosolic Extrinsic Proteins

The covalently linked non-cytosolic extrinsic proteins are located on the external surface of the plasma membrane and remain attached with a help of oligosaccharide to the non-cyto plasmic monolayer.

The transmembrane proteins in the cytosolic and non-cytosolic sides of the plasma membrane can be determined by several ways such as follows: -

A. Water-soluble reagents that are radioactive or fluorescent can be used for labeling the segments lying outside the membrane. The membrane may then be solubilised (isolated) and the proteins are separated by SDS (Sodium Dodecyl Sulphate) and PAGE (poly Acryl amide Gel Electrophoresis) and the labeled proteins are identified. If the proteins from intact cells or sealed ghosts are studied, labeling will suggest external (non-cytosolic) segment and if sealed inside – out ghosts are labeled, then it is internal (cytosolic) segment. If labeling is done by both the cases, it suggests the presence of a transmembrane protein.

B. Intrinsic or Extrinsic Surface of the membrane may be subjected to protease and the isolated proteins may be analyzed. If the protein is partially digested from both the surfaces, it suggests the presence of a transmembrane protein.

C. Specific Antibodies can also be used to find out the location of that specific part of a protein.

4. Non-Covalently Linked Extrinsic Proteins

- These are proteins present on either side of the membrane (Example – Spectrin)
- These are attached to the plasma membrane and other transmembrane proteins by non-covalent interactions
- These are also called as Peripheral membranes (or) extrinsic membrane proteins
- These can be released easily by gentle extraction procedure from the plasma membrane.

5. Intrinsic or Integral Proteins

- All other proteins that cannot be released easily are called as integral or intrinsic proteins
- Integral proteins include transmembrane proteins, covalently linked cytosolic extrinsic proteins and non-cytosolic extrinsic proteins.

6. Representative Membrane Proteins

Spectrin

- Extrinsic, cytosolic cytoskeleton protein, not covalently attached to the membrane
- Most of the membrane proteins in RBC's are peripheral proteins associated with cytosolic side of the membrane
- Among these peripheral proteins, spectrin is the most abundant protein
- Occupies 25 % of the membrane protein mass
- 2.5×10^5 copies per cell
- Spectrin is a thin, long, flexible rod shaped heterodimer protein about 100 nm in length.
- The heterodimer consists of two antiparallel loosely intertwined flexible α and β chains, not covalently attached to each other at several points including both the ends.
- Each of the two α and β chains has repeating domains, 106 amino acids long
- Two heterodimers associate 'head to head' at their phosphorylated heads to form tetramers and four to five tetramers bind short actin filaments with the help of other proteins to form a 'junctional complex'.

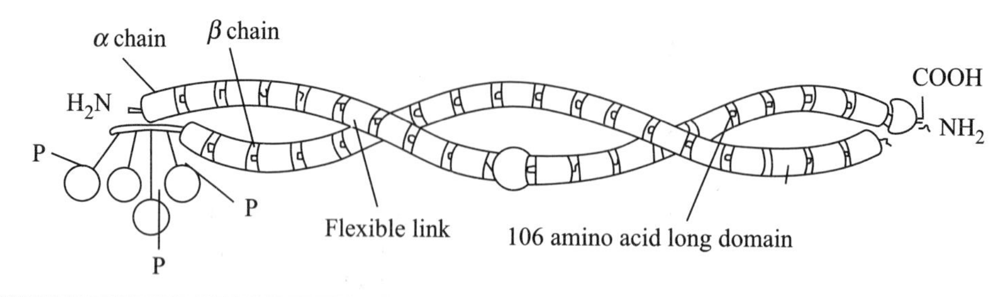

Fig. 3.1 One Spectrin Molecule from Human Red Blood Cell

- Deficiency or mutation in spectrin causes anemia leading to abnormality in the shape of RBC's which become spherical, instead of concave
- Proteins homologous to spectrin are ankyrin and band 4 and these are present in the nucleated cells.

Function

- Maintain structural integrity
- Maintain biconcave shape of the membrane

Glycophorin

- Is a single pass transmembrane protein of α helix with its hydrophilic carboxy-terminal tail exposed to cytosol
- Non-cytosolic external surface – glycophorin carries about 100 sugar residues
- Glycophorin contains 60% of carbohydrates of RBC's
- Nearly 10 lakh (1 million) glycophorin molecules are found per RBC
- Many cell receptors in the nucleated cells belong to the category of glycophorin

Band 3

- Multipass transmembrane protein catalyzing transport of anions (Cl-, HCO_3^-)
- Derived its name from is position relative to other proteins on the PAGE results
- It's a long (930 amino acid residues) polypeptide arranged as dimmers traversing the lipid bilayer as many as 14 times
- There are as many as 10^6 molecules of this protein per erythrocyte

Functions

- Anion transporter
- Helps RBC's to carry CO_2 from tissues to the lungs in the form of HCO_3^- ions in exchange of Cl^- ions

7. Bacteriorhodopsin

- Bacteriorhodopsin is a light activated protein found in the bacteria called Halobacterium halobium
- This bacteria is found in salt-water pools or brine exposed to a large amount of sunlight
- Bacteriorhodopsin is a multipass transmembrane protein having 7 transmembrane α helices which are closely packed
- Each helix has 25 amino acids and these are also called as '7 sister proteins'

- Bacteriorhodopsin is a light activated proton pump
- Bacteriorhodopsin contains a light absorbing group (or) chromatophore (retinal cells) that colors the protein into purple color
- This retinal is related to vitamin A and is identical to chromatophores found in the rhodopsin of the vertebrate eye
- During the activation of light by a photon, the chromatophore changes its shape and causes a series of changes in bacteriorhodopsin protein, resulting in the transfer of one H+ ion from the inside to the outer side of the cell.
- In bright light, several hundred protons can be pumped every second by one bacteriorhodopsin molecule
- This establishes a proton gradient across the membrane facilitating the production of ATP by another protein, thus providing energy to the bacterial cells.

8. Porins

- Porin is a β protein found in the outer membrane of many bacteria like E.coli
- Generally, in the transmembrane proteins, the segments traversing the membrane are α helices and rarely are β helices.
- Porin allows selected hydrophilic solutes to pass across this outer lipid bilayer
- Proteins in the outer membrane of mitochondria and chloroplasts also resemble in having β sheets instead of α helix.
- Porin structure has been studied in Rhodobacter capsulates. It has 16 stranded antiparallel β sheets
- Three monomers associate to form a trimer in porin structure

9. Protein Complex in Membrane

- Sometimes, several proteins associate together in the membrane forming protein complexes performing a variety of complex function
- One such complex is 'photosynthetic reaction centre'
- This is studied in detail in the bacteria, Rhodopseudomonas viridis
- The structure consists of four subunits namely L, M, H and a cytochrome
- L and M have five α helices and form the core of the reaction center
- Contains electron carrier co-enzymes
- These type of protein complexes are formed by a variety of proteins that harvest energy and transduce signals from the outside world into intracellular pathway

10. Specific Proteins for Specific Membrane Domains

- Although proteins are mobile within the lipid bilayer, specific proteins are there that perform specific functions and confine themselves to that particular domains
- Example – Cells lining the gut and kidney tubules.
- There are specific proteins confined to the apical domain and to the basolateral domain
- There are also mechanisms to transport these specific proteins to specific domains within the cells

Cell coat

- Also called as Glycocalyx (rich in sugar residues)
- Most of the lipids in the bilayer and non-cytosolic external segments of the extrinsic as well as intrinsic proteins of the plasma membrane are glycosylated and are therefore called as glycolipids and glycoproteins
- This is true in both plant and animal cells
- All these sugar rich molecules make a sugar rich zone on the surface of plasma membrane called as cell coat or glycocalyx

Function

- Protection, cell recognition, cell-cell adhesion and inflammatory responses

CAMs (Cell Adhesion Molecules)

- During development of higher organisms, differentiation and morphogenesis occur
- Cell recognition and adhesion takes place between cell-cell and cell-substratum etc
- This cell adhesion is mediated by CAMs, substrate adhesion molecules and cell junction molecules
- These molecules (excluding cell junction) have four major receptors

Cadherins

- Cadherins mediate cell adhesion, helps in normal development, and maintains the integrity of multicellular structures and cell-cell contact.
- Cadherins depends on calcium for their function and removal of calcium abolishes adhesive activity. They act both as a receptor and a ligand

Integrins

- Integrins involves cell–substratum adhesion and have been found to relay messages from the extracellular matrix into the cell.
- They help to integrate many of the diverse signals that impinge on the cells and thereby determine the cell's fate.

Selectins

- Selectins are multifunctional cell-cell adhesion molecules expressed on the surface of leukocytes and activated endothelial cells.
- Facilitate inflammatory responses and plays a major role in signalling process.

N-CAM

- Neural cell-adhesion molecule (N-CAM) plays a major role in cell-cell adhesion, neurite outgrowth, synaptic plasticity, and learning and memory.
- N-CAM is a homophilic binding glycoprotein expressed on the surface of neurons, glia and skeletal muscles.

3.2 CYTOSKELETAL PROTEINS

A network of protein filaments known as cytoskeleton spatially organizes the cytoplasm of eukaryotic cells. This network contains three principal types of filaments namely microtubules, actin filaments and intermediate filaments. Microtubules are stiff structures that usually have one end anchored in the centrosome and the other end free in the cytoplasm. In many cells, microtubules are highly dynamic structures that alternately grow and shrink by the addition and loss of tubulin subunits. Motor proteins move in one direction or the other along microtubules carrying specific membrane bounded organelles to desired locations in the cells. Actin filaments are also dynamic structures but they normally exist in bundles or networks rather than as single filament. A layer called the cortex is formed just beneath the plasma membrane from actin filaments and a variety of actin binding proteins. This actin-rich layer controls the shape and surface movements of most animal cells. Intermediate filaments are relatively tough, rope like structures that provide mechanical stability to cells and tissues. The three types of filaments are connected to one another and their functions are coordinated.

Definition

Cytoskeleton is a structure associated with the plasma membrane consisting of three elements such as **microtubules, actin filaments** or **microfilaments** and **intermediate filaments**.

The important functions of cytoskeleton are to provide stability to the cell shape and control of cell movement between and within the cells.

1. Microtubules

Definition

- Microtubules are long protein fibres lying mainly in the ectoplasm of cell cortex. i.e. in the cytoplasm below the plasma membrane
- 20-30 nm in diameter

Occurrence

- These are regular organelles in most animal and plant cells except amoeba and mature mammalian erythrocytes
- These occur in a variety of cell structures including cilia and flagella, centrioles and basal bodies, nerve processes, mitotic apparatus, cell cortex in meristematic cells etc.

Structure

- Microtubules are cylindrical structures, branched and are several microns in length (20-30 nm)
- In transverse section, these microtubules appear to consist of a circular array of 13 subunits and in surface view, a microtubule appears to consist
- Of 13 rows of subunits. These subunits are globular in shape and are called protofilaments7

Chemical composition

- Each subunit has a sedimentation coefficient of 6S and two globular polypeptides namely α **tubulin** and β **tubulin**, each with a molecular weight of 60000 Daltons and an amino acid similar to that of a muscle protein called actin or a microfilament
- Association of α **tubulin** and β **tubulin** forms a **dimeric protein** called **tubulin**

Proteins associated with tubulin or microtubules

- Dynein 1: ATPase activity

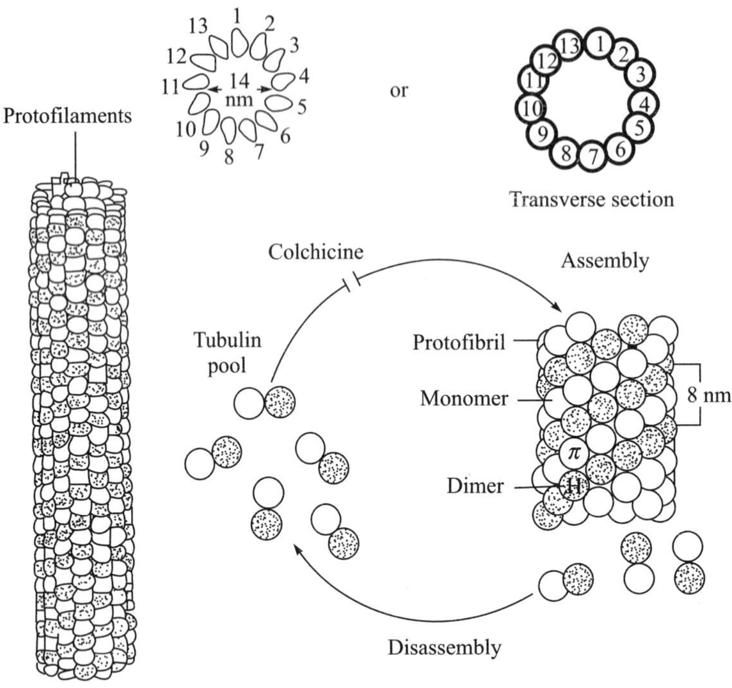

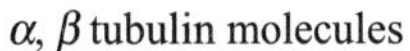

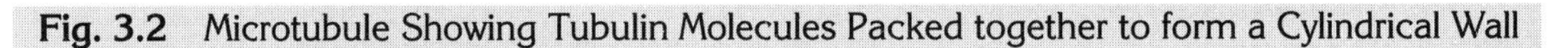

Fig. 3.2 Microtubule Showing Tubulin Molecules Packed together to form a Cylindrical Wall

- Dynein 2: Generation of movement
- Nexin: forms linkages
- MAP 1: high molecular weight proteins – initiation and elongation of microtubules
- MAP 2: High molecular weight proteins – initiation and elongation of microtubules
- Tau: promotes assembly of tubulin molecules in rings and microtubules
- Kinase: phosphorylation of tubulin subunits
- Kinesins and Kinectin: Motor proteins in membrane traffic

2. Microfilaments or Actin Filaments

Definition

Microfilaments otherwise called as Actin filaments are long fibres of variable length each with a diameter of 5-7 nm and consist of protein molecules called actin

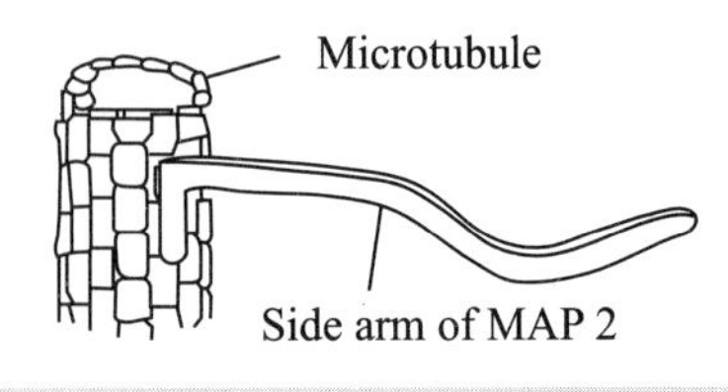

Fig. 3.3 Structure of a Microtubule

Occurrence

They occur as rod shaped structures forming a web in the ectoplasm or cell cortex. The region of the cell, crowded with actin filaments is often devoid of cell organelles such as plastids, mitochondria, endoplasmic reticulum, Golgi complex, ribosomes etc. These were initially studied in muscle cells but are now found in all types of cells.

Functions

- Movement: formation of cell furrow or cell plate after cell division
- Cytoplasmic streaming
- Cell migration during early stages of embryo development
- All these processes are inhibited if actin filaments are destroyed by an antibiotic cytochalasin B.

Actin filaments in striated muscles

- Example: Biceps of the upper arm
- It consists of a large number of muscle fibres
- Each fibre is a cell with several nuclei and is made up of long strands of 1-2 micrometer in diameter and these strands are called myofibrils
- Each myofibril consist of a chain of units called sarcomeres
- Sarcomeres are lined adjacent to myofibrils giving a striated appearance to the muscle. Sarcomeres are long and are separated from each other by dark **Z lines or Z discs**. Between two successive Z-Lines a sarcomere is divided into **I bands** (light in color) and **A bands** (Dark in color)
- Each sarcomere has two sets of parallel overlapping filaments called thick filaments and thin filaments

Actin and myosin proteins

- Thin filaments: actin protein
- Thick filaments: Myosin protein
- Actin protein has a subunit called G-actin that is globular and 5nm in diameter and polymerizes to form filamentous or F-actin
- Myosin proteins have a long tail with two globular heads
- The tail units of different myosin molecules overlap in the middle of the thick filaments making cross-bridges with the thick and thin filaments
- During these connections, ATPase activity associated with myosin heads is stimulated by actin protein and each myosin molecule hydrolyses 5-10 molecules of ATP per second
- Body stiffness on death results due to lack of detachment of myosin head from actin due to non-availability of ATP, which is essential for detachment. This stage is called **rigor mortis**.

Actin filaments in smooth muscles

- Found in the cytoplasm

Functions

- Cytoplasmic streaming
- Cell locomotion (Pseudopodia in amoeba)

3. Intermediate Filaments (IF)

Definition

Intermediate Filaments are filaments that range in size between microtubules (20-30 nm) and actin filaments (5-7 nm). These intermediate filaments were discovered in a protein called keratin, found in hairs.

Occurrence

- Makes 1% of total protein in all cells
- There are 50 different IF filament genes which are differentially expressed in different types of cell

Structure

Each basic unit of an IF has a dimmer composed of two α - helical chains, oriented in

A parallel way and interwined in a coiled-coil rod and there are four dimmers totally. These IF dimmers have a polar coiled side and apolar side.

Unique properties

- IF's of hairs and epidermal cells are highly insoluble
- IF's of nuclear lamina lining the inner surface of the nuclear membrane are soluble
- Fibroblasts are soluble
- Non-helical heads and tails

Types of intermediate filaments (IF's)

There are four types of intermediate filaments based on the morphology and localization within the cell.

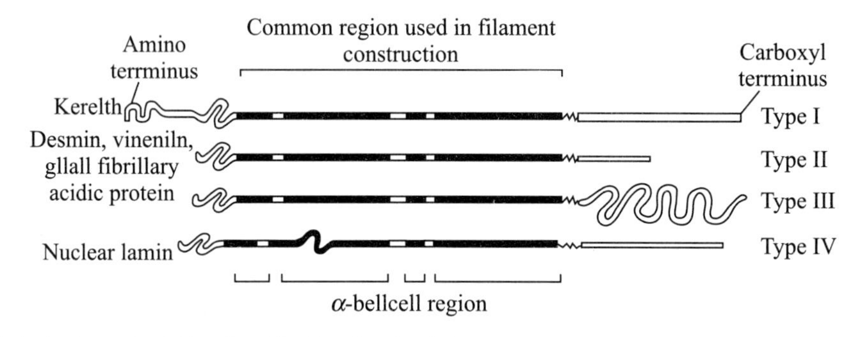

Fig. 3.4 Intermediate Filaments Share a Similar Central Region with Three Short Interruptions

1. Keratin filaments

- Also known as tonofilaments, pre-keratin and cyto-keratin
- Found in epithelial cells
- Mammalian cytokeratins are composed of polypeptides ranging in size from 47000 to 58000 Daltons
- They are α-fibrous proteins that form bulk of the dead layers or stratum corneum

2. Neurofilaments

- Neurofilaments together with microtubules form the main constituents of axons, dendrites and neuronal perikaryon
- They have 3 polypeptides ranging in molecular weight from 20000 – 68000 Daltons

3. Glial filaments

- Found throughout the cytoplasm of astrocytes
- Composed of a very acidic protein, 51000 Daltons in molecular weight

4. Heterogenous filaments

- Consist different proteins such as Desmin, Vimentin and Synemin
- Desmin is found in the smooth, skeletal and cardiac muscle
- Viementin is found in different cell origins
- Synemin is found in skeletal muscles

3.3 EXTRA CELLULAR MATRIX

Tissues are not made solely of cells. A substantial part of their volume is *extracellular space*, which is filled by a network of macromolecules constituting the **extracellular matrix**.

This extracellular matrix is composed of a variety of proteins and polysaccharides that are secreted locally and assembled into an organized meshwork in closed association with the surface of the cell that produced them. The extracellular matrix concentrates chiefly on the connective tissues and in these tissues, the matrix is more plentiful than the cells that it surrounds and it determines the tissues physical properties. The extra cellular matrix found in different parts or organs vary greatly from skin and bone and to brain and spinal cord.

Some **characteristic features** of extra-cellular matrix are as follows: -

1. The extracellular matrix is made and oriented by the cells within it
2. Glycosaminoglycan (GAG) chains occupy large amounts of space and form hydrated gels. GAG's are unbranched polysaccharide chains composed of repeating disaccharide units. They are called GAG's because one of the two sugar residues in the repeating disaccharide is always an amino sugar (N-acetylglucosamine or N-acetylgalactosamine), which in most cases is sulfated. The second sugar is usually an uronic acid (glucuronic or iduronic). Since there are sulfate or carboxyl groups on most of the sugar residues, GAG's are highly charged. Four main groups of GAG's have been recognized so far and they are (1) hyaluronan, (2) chondroitin sulfate and dermatan sulfate, (3) heparan sulfate and (4) keratan sulfate.
3. Hyaluronan is thought to facilitate cell migration during tissue morphogenesis and repair. Hyaluronan is also called as hyaluronic acid or hyaluronate. This is one of the simplest GAGs. It consists of a regular repeating sequence of upto 25000 nonsulfated disaccharide units. It is found in variable amounts in all tissues and fluids in adult animals and is especially abundant in early embryos. This plays a major role in resisting compressive forces in tissues and joints. This also plays a major role in space filling during embryonic development.
4. Proteoglycans are composed of GAG chains that are covalently linked to a core protein. Except hyaluronan, all GAGs are found covalently attached to protein in the form of proteoglycans that constitute majority of the animal cells.
5. Proteoglycans can regulate the activities of secreted signalling molecules and these GAGs can be highly organized in the extra cellular matrix. Cell surface proteoglycans act as co-receptors also.
6. Collagens are the major proteins of the extra cellular matrix

3.4 CELL DIVISION (MITOSIS & MEIOSIS)

Cells reproduce by duplicating their contents and then dividing into two. This cell division cycle is the fundamental means by which all living things are propagated. In unicellular species, such as bacteria and yeasts, each cell division produces an additional organism. In multicellular species, many rounds of cell division are required to make a new individual and cell division is needed in the adult too, to replace cells that are lost by wear and tear or by programmed cell death. The duration of the cell cycle varies greatly from one cell type to another. The cell cycle is traditionally divided into several distinct phases of which the most dramatic is mitosis, the process of nuclear division, leading to the cell division.

There are two kinds of phases in cell division and they are namely, Mitosis and Meiosis.

1. Mitosis

Mitosis is a phase of cell division in which the development of an individual from zygote to an adult stage takes place. In Mitosis, multiplication of cell number takes place and it's a continuous process-giving rise to two daughter cells that resemble each other and also the parent cell qualitatively and quantitatively.

The mitosis phase consists of the following stages:

1. **P**rophase
2. Metaphase
3. Anaphase
4. Telophase
5. Cytokinesis

1. Prophase

- **Cell** gets prepared for cell division
- At the beginning of prophase, chromosomes appear as thin, filamentous uncoiled structures
- In the middle stage, the chromosomes appear coiled, shortened and more distinct
- In the late prophase, longitudinal splitting of each chromosome into two sister chromatids are visible prominently.
- Double structure of each chromosome are seen clearly
- Sister chromatids at this stage are attached only at the centromere
- The two chromatids remain attached to the spindle tubules with the help of kinetochores
- Nuclear membrane disappears
- Nucleolus also disappears

2. Metaphase

- During late prophase and early metaphase, spindle fibres start appearing.
- These tubules get attached to the chromosomes at centromeres with the help of kinetochores.
- The chromosomes arrange themselves in the equatorial plate.
- The spindle apparatus that helps the chromosomes to arrange themselves in the equatorial plane is formed with the help of centrosome particularly in the case of animal cells.
- Centrosomes have two centrioles, which spare at the time of spindle formation and lie on opposite sides of the nucleus.
- When the two centrioles separate, astral rays start radiating outward from each centriole.
- These will join and form spindle fibres.
- The characteristic feature of this phase is to determine the number of chromosomes.

3. Anaphase

- After spindle is formed and the chromosomes get arranged on the plane, the chromosomes split at the centromeres also.
- Sister centromeres separate from each other and are called daughter chromosomes.
- These daughter chromosomes now move towards opposite poles of the spindle.
- Movement is by contraction of spindle fibres.

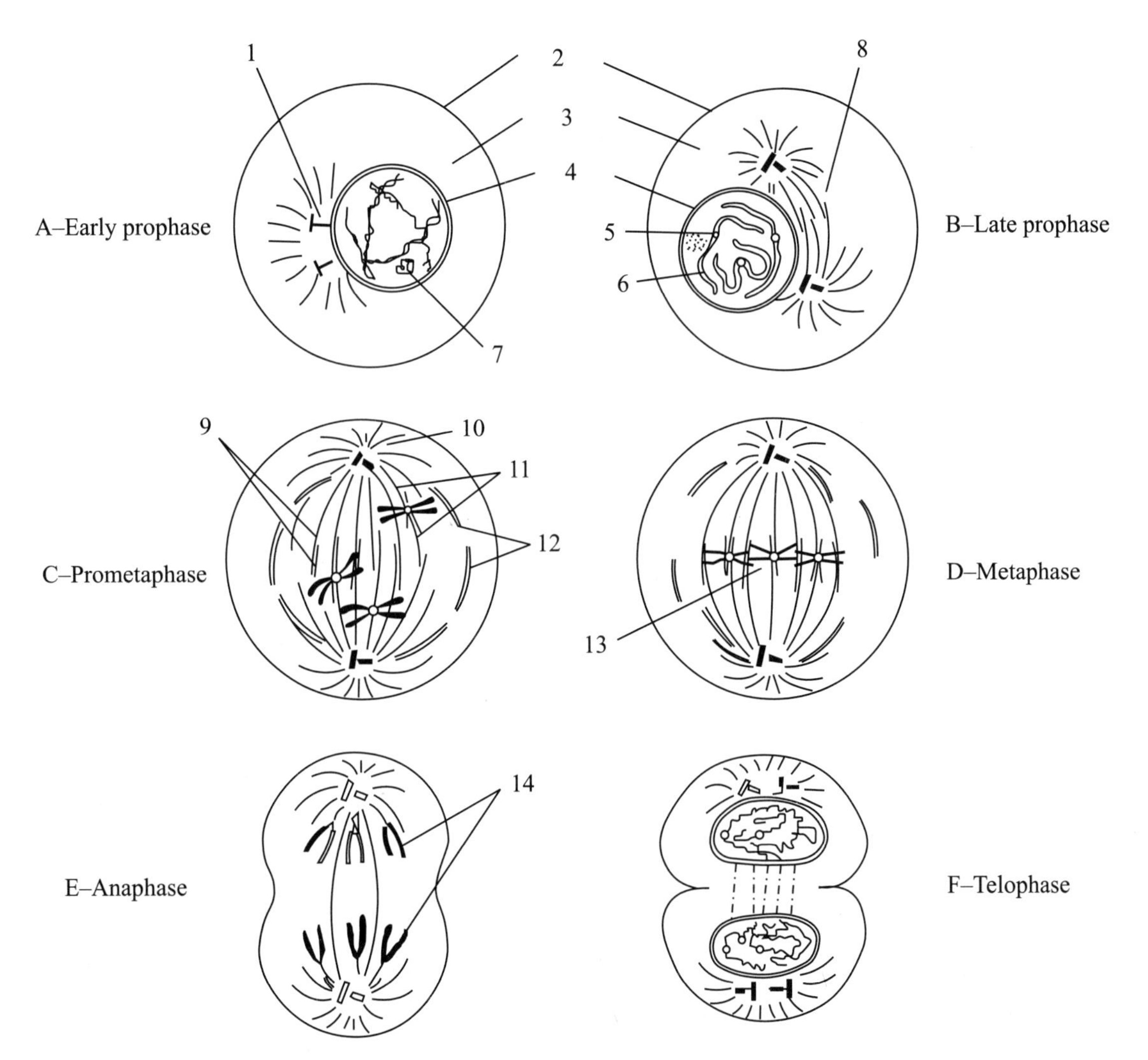

1-centrioles, 2-plasma membrane, 3-cytoplasm, 4-nuclear membrane, 5-centromere, 6-condensing chromosomes, 7-nucleolus, 8-developing bipolar spindle, 9-polar microtubules, 10-spindle poles, 11-kinetochore microtubules, 12-nuclear membrane fragments, 13-stationary chromosomes aligned at metaphase plate (equator), 14-chromosomes

Fig. 3.5 Mitosis in an Animal Cell (After Burns and Bottino, 1989)

4. Telophase

- Nuclear membrane is re-constructed around each group of chromosome giving rise to one nucleus at each pole
- Nucleoli starts reappearing and this process is called nucleologenesis

5. Cytokinesis

- Division of one nucleus into two is called as karyokinesis and it is followed by cytokinesis that means division of cytoplasm into two cells.
- Division of cytoplasm takes place by two ways: -
- *Cell furrow*: - Ex. animal cells. In case of animals, outer layers are more flexible due to the absence of cell wall. In such cases, a circular constriction appears at the equator and it converges on all sides finally separating two daughter cells.
- *Cell Plate*: - ex. Plant cells. In plant cells, a more rigid cell plate is usually initiated at the center and is completed towards the periphery

The entire process of mitosis takes 10 minutes to several hours. The number of chromosomes in some important plants and animals are as follows: -

Human beings – 46 (Diploid), Rat – 42, Rabbit – 44Drosophila melanogaster – 8, Rice – 24, Wheat – 42, Garden pea – 14

Significance of Mitosis

- Helps the cell in the maintenance of size
- Helps in the maintenance of equilibrium in the amount of DNA and RNA in the cell
- Provides an opportunity for the growth and development to organs and the body of the organisms
- The old decaying and dead cells are replaced by new ones by the process of mitosis
- Gonads and the sex cells depend on the mitosis for the increase in their number
- Cleavage of egg during embryo genesis and division of blastema during blastogenesis involves mitosis.

2. Meiosis

Meiotic division consists of two successive divisions of the cell and they are namely, First Meiotic division and Second Meiotic division.

The first meiotic division is otherwise called as heterotypic division. Here, reduction of chromosome number takes place and thus two haploid cells are formed in this division.

The second meiotic division is otherwise called as homotypic cell division and in this division, the haploid cells divides mitotically and results into four haploid cells.

I Meiotic division

- This is a reduction division
- First meiotic division consists of
  - (a) Prophase I
  - (b) Metaphase I

(c) Anaphase I
(d) Telophase I

(a) Prophase I

- This is a complex phase and is of a very long duration process and most useful process that include cytogenetical events such as synapsis and crossing over etc.
- This is the longest meiotic phase and differs from mitotic prophase in several ways and therefore, for convenience are divided further into five stages namely,

  i. Leptotene
  ii zygotene
  iii. Pachytene
  iv. Diplotene
  v. Diakinesis

i. Leptotene

- First stage of meiosis following the interphase
- Chromosomes at this stage appear as long thread like structures that are loosely woven
- On these thread like chromosomes, bead like structures are found and they are called the chromomeres
- No chromatin material is seen
- The centrioles duplicate and migrate towards the poles

ii. Zygotene

- Pairing of homologous chromosomes and this process is otherwise called Synapsis
- The synapsis begins at one or more points along the length of the homologous chromosome. Three types of synapsis take place and they are 1. proterminal synapsis, 2. procentric synapsis and 3. Random synapsis.
- A thick, protein-containing network called synaptonemal complex joins the paired homologous chromosomes.

iii. Pachytene

- The pair of chromosomes become twisted spirally around each other and cannot be distinguished separately.
- In the middle of this pachytene stage, each homologous chromosome splits length-wise to form two chromatids.
- During pachytene stage, an important genetic phenomenon called 'crossing over' takes place. The crossing over involves reshuffling, redistribution, and mutual exchange of hereditary material of two parents between two homologous chromosomes.
- The two chromatids are still linked to their common centromere. And this synaptonemal pair is referred to as a bivalent because of its visible two chromosomes, or as a tetrad because of the four visible chromatids.
- After the division of chromatids, the interchange of chromatid segments takes place between the non-sister chromatids of the homologous chromosomes.
- An enzyme called ligase unites the broken chromatid segments.

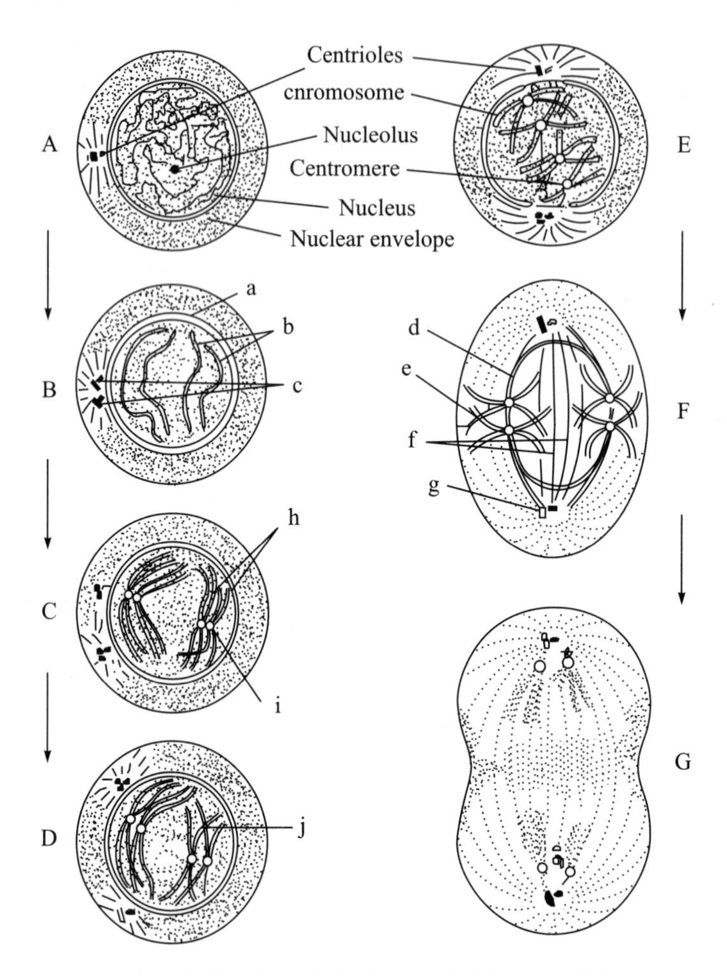

A–Leptotene, B–Zygotene, C–Pachytene, D–Diplotene, E–Diakinesis, F–Metaphase and G–Late metaphase

a-nuclear envelope, b-homologous chromosome, c-daughter centriole, d-chromosomal fibres, e-chiasmata, f-continuos fibres, g-centrioles, h-chromatids, i-centromere, j-chiasmata

Fig. 3.6 Different stages of First Meiotic Division (after King 1965)

- The process of interchange of chromatin material between one non-sister chromatid of each homologous chromosome is known as 'crossing over'
- Crossing over is followed by chiasmata formation
- Synthesis of small amount of DNA takes place and this is utilized during crossing over and chiasmata formation

iv. Diplotene

- The chromatids of each tetrad are clearly visible

- The synaptonemal complex appears to be dissolved leaving participating chromatids of the paired homologous chromosomes physically joined at one or more discrete points called chiasmata.
- At these points, crossing over takes place.
- Unfolding of chromatids takes place facilitating RNA synthesis and cell growth

v. Diakinesis

- The bivalent chromosomes becomes more condensed and are evenly distributed in the nucleus
- The nucleolus disappears
- The nuclear envelope breaks down
- The chiasma moves from the centromere towards the ends of the chromosomes and the intermediate chiasmatas disappear
- This movement of chiasmata is called terminalization.

(b) Metaphase I

- Metaphase consists of a prometaphase stage in which the nuclear envelope disintegrates and the microtubules get arranged in the form of spindle in between the two centrioles that occupy the position of two opposite poles of the cell.
- Spindle fibres get attached to the chromosomes and chromosomes get aligned themselves at the equator.
- The microtubules of the spindle get attached to the centromere of the homologous chromosomes of each tetrad.
- The centromere of each tetrad is directed towards the opposite poles.
- The repulsive forces between the homologous chromosomes increases greatly and the chromosomes become ready to separate

(c) Anaphase I

- Reduction (actual) takes place at this phase.
- The homologous chromosomes, which move towards the opposite poles, are the chromosomes of either paternal or maternal in origin.
- During the chiasma formation, out of two chromatids of a chromosome, one has changed its counterpart and therefore, two chromatids of a chromosome do not resemble with each other in genetic terms.

(d) Telophase I

- Arrival of a haploid set of chromosomes at each pole defines Telophase I.
- Nuclei and nucleolus re-appears.
- Two daughter chromosomes are formed.
- After karyokinesis, cytokinesis occurs and two daughter cells with haploid chromosomes are formed.

Both cells undergo a resting phase for a short time and this is called as Interphase. During interphase, no DNA replication occurs.

II Meiotic division

The second meiotic division otherwise called as the homotypic division is actually the second mitotic division that divides each haploid meiotic cell into two haploid cells. The second meiotic division includes the following four stages.

(a) Prophase II
(b) Metaphase II
(c) Anaphase II
(d) Telophase II

(a) Prophase II

- Each centriole divides into two and thus two pairs of centrioles are formed.
- Each pair of centrioles migrate towards the opposite poles
- The microtubules get arranged in the form of spindle at the right angle of the spindle of first meiosis.
- The nuclear membrane and nucleolus disappear
- The chromosomes with two chromatids become short and thick

(b) Metaphase II

- The chromosomes get arranged on the equator of the spindle

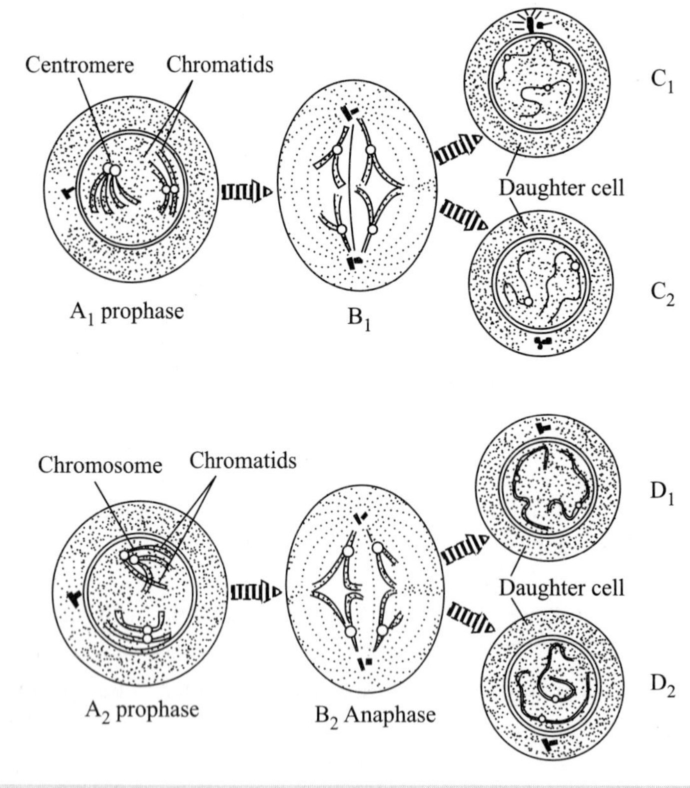

Fig. 3.7 Second Meiotic Division (After King, 1965)

- The centromere divides into two daughter chromosomes
- The microtubules of the spindle are attached with the centromere of the chromosome

(c) Anaphase II

- The daughter chromosomes move towards the opposite poles due to the shortening of chromosomal microtubules
- Stretching of interzonal microtubules of the spindle

(c) Telophase II

- Chromatids migrate towards the opposite poles and are now known as chromosomes
- Nuclear envelope is formed around the chromosomes
- Nucleolus re-appears due to the synthesis of ribosomal RNA and also due to accumulation of ribosomal proteins.
- After the karokinesis, in each haploid meiotic cell, the cytokinesis occurs resulting in the formation of four haploid daughter cells.
- These cells have different types of chromosomes due to the crossing over in the prophase I.

Significance of Meiosis

- The meiosis maintains a definite and constant number of chromosomes in the organisms
- By crossing over, meiosis provides an opportunity for the exchange of the genes and thus causes genetic variations among species
- These variations are the raw materials for the process of evolution

3.5 CELL CYCLE AND MOLECULES THAT CONTROL CELL CYCLE

The complete cell division consists of five stages namely

1. G_1
2. S
3. G_2
4. M
5. D

The first three stages together constitute the Interphase. Interphase is the period of rest between cell divisions. M stage is the process of cell division till the formation of nuclei in the daughter cell and this is referred to the mitotic division and therefore called as 'M' stage. The remaining activities till cytokinesis are termed as 'D' and refer to Division.

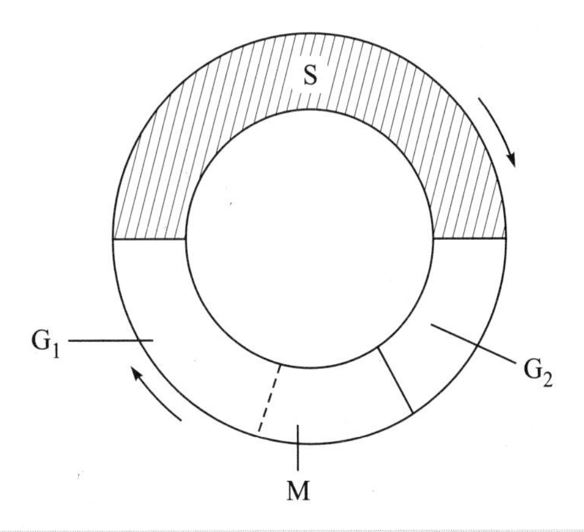

Fig. 3.8 The Cell Cycle Showing the Four Phases

- During the interphase, DNA, RNA and protein are synthesized
- Metabolic activity is greatest during this period

Table 1 Comparison between Mitosis and Meiosis

	Mitosis	*Meiosis*
1.	Mitosis occurs continuously in the body or somatic cells	Meiosis occurs in the germ cells (testes or ovaries)
2.	The whole process gets completed in one phase	The whole process gets completed in two successive divisions, one after the other
		Prophase
3.	The prophase is of short duration and has no sub-stages	The prophase is of longer duration and has five successive stages namely, Leptotene, Zygotene, Pachytene, Diplotene and Diakinesis
4.	The homologous chromosomes duplicate into two chromatids. The two chromatids separate and form two new chromosomes. Each daughter cell receives the chromatids of each homologous chromosome and thus has the exact chromosome number like the parental cells.	Out of two homologous chromosomes, only one type of chromosome either maternal or paternal moves to the daughter cells. A daughter cell thus receives only a paternal or maternal move to the daughter cells. A daughter cell thus receives only a maternal or paternal chromosome of the homologous pair and the number of chromosomes remains half than the paternal cells.
5.	No pairing or synapsis takes place between the homologous chromosomes	Pairing or synapsis occurs between the homologous chromosomes
6.	Duplication of chromosomes takes place in the early prophase	Duplication or splitting of chromosomes takes place in the late prophase (Pachytene stage)
7.	No chiasma formation or crossing over takes place	Chiasma formation or crossing over takes place
8.	Exchange of the genetic material between the homologous chromosomes does not occur	Exchange of the genetic material takes place between the non-sister chromatids of homologous chromosomes
		Metaphase
9.	The chromatids occur in the form of dyads	The chromatids of two homologous chromosomes occur as a tetrad
10.	The centromere of the chromosome remain directed towards the equator and the arms of the chromosome remain directed towards the pole	The centromere of the chromosome remain directed towards the poles and the arms of the chromosome remain directed towards the equator
		Anaphase
11.	The chromosomes are the monads (having single chromatid)	The chromosomes are the biads (having two chromatids and single centromere)
12.	Chromosomes are long and thin	Chromosomes are short and thick
		Telophase
13.	Always occurs	Fist telophase is sometimes omitted
		Significance
14.	The chromosome number in each daughter cell remains the same like the parent cell	The chromosome number is reduced to half in the daughter cells than the parental cells
15.	A diploid cell produces four diploid cells by the mitotic division	A diploid cell produces four haploid cells by a meiotic division

- Different eukaryotic cells vary in the length of time taken to complete an entire life cycle
- They also vary in the relative proportions allotted to each of the four stages.
- Therefore, in continuously dividing cells, an individual cell passes through the following main phases of cell cycle called as
  1. Interphase (G_1, S and G_2)
  2. Mitotic Phase (Karyokinesis)
  3. Cytokinesis (cell division)

Interphase

- The resting phase or stage between two mitotic divisions is called interphase or intermitotic phase.
- During Interphase, no division of chromosomes or cytoplasm occurs but the nucleus and cytoplasm remain metabolically active and due to that an increase in volume of nuclear as well as cytoplasmic substances takes place.
- The interphase is the longest stage in the cell cycle and takes one or two days for its completion and includes three sub-phases.
- G1 – includes synthesis and organization of substrate and enzyme for DNA synthesis. Includes transcription of rRNA, tRNA, mRNA and synthesis of proteins.
- S phase: chromosomal as well as DNA replication.
- G2: post-DNA synthesis phase – growth of cytoplasm, organelles and macromolecules.
- Nucleus remains intact but chromosomes become diffused, long, coiled and indistinctly visible chromatic fibres are seen.

Different eukaryotic cells vary in duration (minutes) taken to complete an entire cell cycle.

Cell	*Prophase*	*Metaphase*	*Anaphase*	*Teolophase*
Onion root	71	65	2.4	3.8
Mouse spleen	20-35	6-15	8-14	9-26
Pea endosperm	18	17-38	14-26	28

Questions

1. Define membrane proteins.
2. What are transmembrane Proteins?
3. Write about Covalently linked Cytosolic Extrinsic Proteins.
4. Define and explain Covalently linked Non-cytosolic Extrinsic Proteins.
5. What are Non-Covalently linked Extrinsic Proteins?
6. Define Integral or Intrinsic Proteins.
7. Mention a few Representative Membrane Proteins.
8. Describe the structure of Spectrin.
9. Enumerate two significant characteristics of Glycophorin.
10. Describe Band 3.

11. Describe Bacteriorhodopsin.
12. What are Porins?
13. What are the Protein Complexes in membranes?
14. Are there any specific Proteins for Specific Membrane Domains? If so, what are they?
15. Describe in brief about the cell coat.
16. What are CAMs (Cell Adhesion Molecules)?
17. Define Cadherins.
18. Define and describe the structure of Integrins.
19. Describe the structure and use of Selectins.
20. What do you know about N?-CAM.
21. Describe hydropathy plot.
22. What are Microtubules?
23. Write about actin filaments.
24. Describe Intermediate Filaments.
25. What are the proteins associated with tubulin?
26. What are keratin and glial filaments?
27. Define mitosis and write the stages of mitosis?
28. Define meiosis and write the stages that occur under the first and second meiotic division.
29. Explain the cell cycle and explain the stages that occur in a normal cell.
30. Explain the S Phase in a cell cycle.

CHAPTER 4

Internal Molecular Organization of the Cell

4.1 PRINCIPLES OF MEMBRANE ORGANIZATION

Biological membranes consist of a continuous double layer of lipid molecules in which various membrane proteins are embedded. This lipid bilayer is fluid, with individual lipid molecules able to diffuse rapidly within their own monolayer. Most types of lipid molecules however, very rarely, flip-flop spontaneously fro m one monolayer to the other. Membrane lipid molecules are amphipathic and some of them (the phospholipids) assemble spontaneously into bilayers when placed in water; the bilayer from sealed compartments that reseal, if torn. There are three major classes of membrane lipid molecules – phospholipids, cholesterol and glycoplipids – and the lipid composition of the inner and outer monolayers are different, reflecting the different functions of the two faces of a cell membrane. Different mixtures of lipids are found in the membranes of cells of different types as well as in the various membranes of a single eukatyotic cell. Some membrane bound proteins require specific lipid head groups in order to function that contain many kinds of lipid molecules.

Both, prokaryotic and eukaryotic cells are enclosed by a plasma membrane that physically separates the cytoplasm from the surrounding cellular environment. Plasma membrane is a ultrathin, elastic, living dynamic and selective transport-barrier. It is a fluid mosaic assembly of molecules of lipids (phospholipids

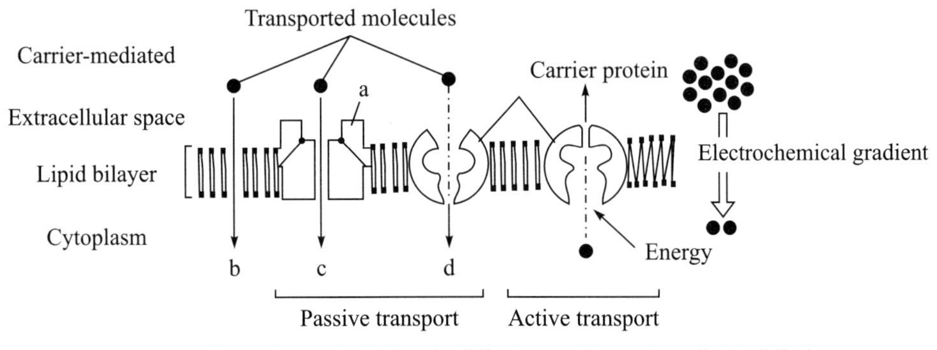

a-Channel protein, b-Simple diffusion, c- Channel-mediated diffusion

Fig. 4.1 Types of Transport Across the Membranes. (After Alberts, *et al.*, 1989)

and cholesterol), proteins and carbohydrates. Plasma membrane controls the entry and exit of nutrients, waste products and maintains the differences in ion concentration between the interior and exterior of the cell. Plasma membrane senses external signals such as hormones or immunological and responds to environmental signals too. All the biological membranes that surround eukaryotic cell organelles such as mitochondria, chloroplast, nucleus, Golgi apparatus, lysosomes, and peroxisomes etc., including plasma membrane are similar in structure and vary in functions in having permeability.

C.Nageli and C.Cramer coined the term 'plasma membrane' in the year 1855 and the plasma membrane is also called as cytoplasmic membrane, cell membrane or plasmalemma. J.Q.Plowe coined the term plasmalemma in the year 1931.

Study of Plasma Membrane

The structure of the plasma membrane of various cells has been studied by isolating them from living systems and also by their artificial synthesis by using their constituent molecules. The pure and isolated membranes are then studied by biochemical and biophysical methods. The studies of plasma membrane structure and constituents are carried out using electron microscopy, enzyme analysis and topology of surface antigens. Samples of plasma membrane were taken from mammalian erythrocytes and the myelin sheath of the nerve fibre. The erythrocytes or the red blood cells were taken for experiments because they have the advantages such as (a) easily obtained (b) these cells do not contain any intracellular organelles or membrane (c) the erythrocyte membrane is relatively tough in a way that it does not get easily fragmented.

Plasma menbrane are isolated from erythrocytes by a process called as haemolysis. Haemolyis is a process in which the cells are treated with hypotonic solutions that cause the cell to swell and then loss the haemoglobin content. The resulting membrane is called as a red cell ghost. If the haemolyis process is mild, the permeability functions of the membrane can be restored by certain treatments and such a treated membrane is called as a resealed ghost. When haemolysis is more drastic and there are no chances of resealing the membrane, the resulting membrane is called as a white ghost. The resealed ghosts are used to study the physiological and the biochemical properties of the membrane whereas the white ghosts are used to study only the biochemical properties of the membrane.

Table 1 Chemical Composition of the Plasma Membrane

	Membrane	*Protein*	*Lipid*	*Carbohydrates*
1.	Myelin (nerve cell)	18	79	3
2.	Mouse liver	44	52	4
3.	Amoeba	54	42	4
4.	Human erythrocytes	52	40	8
5.	Inner membrane of mitochondria	76	24	0

Lipids

There are four major classes of lipids in the plasma membrane and other membranes. They are phospholipids, sphingolipids, glycolipids and sterols. All of these four lipids are amphipathic molecules, possessing both hydrophilic and hydrophobic domains and vary in their proportions. The phospholipids may be acidic such as sphingomyelin or neutral phospholipids such as phosphatidyl choline, phosphatidyl serine etc. Another component that is present abundantly is the cholesterol and mammalian cells contain more and it is totally absent in prokaryotic cells.

Proteins

There are two kinds of proteins called as intrinsic proteins or integral proteins and the extrinsic proteins or peripheral proteins. These two may be ectoproteins lying externally to the extracytoplasmic surface of the plasma membrane. The intrinsic proteins tend to affirm closely with the membrane, while the extrinsic proteins have a weaker association and are bound to lipids of membrane by electrostatic interaction. On their basis of their functions, proteins of plasma membrane can also be classified into three types. They are structural proteins, enzymes and transport proteins (permeases or carriers).

Carbohydrates

Carbohydrates are present only in the plasma membrane. They are present as short, unbranched or branched chains of sugars (oligosaccharides) attached either to exterior ectoproteins (glycoproteins) or to the polar ends of the phospholipids at the external surface of the plasma membrane (glycolipids). No carbohydrate is located at the cytoplasmic or inner surface of the plasma membrane. The various combinations of six principal sugars such as D-galactose, D-mannose, L-fructose, N-acetylneuraminic acid, N-acetyl D-glucosamine and N-acetyl D-galactosamine, form all types of oligosaccharides of the plasma membrane.

Enzymes

Some important enzymes present in the plasma membrane are Acetyl phosphatase, acetyl cholinesterase, acid phosphatase, adenosine triphosphatases, RNAase, alkaline phosphatases, amino peptidases, lactase, maltase, sialidase and protein kinase.

4.2 TRANSPORT ACROSS CELL MEMBRANE

Passive Transport (Passive Diffusion)

Transport of small uncharged molar molecules

- Passive transport is defined as transport of solutes from a region of higher concentration to the region of lower concentration till both the concentrations becomes the same.
- This process rarely uses carrier proteins
- Example : - Passive glucose transporters of mammalian cells

The free energy (or) chemical potential difference $=\Delta G$

Rate of Flow $=$ Jc (uncharged molecule)

Are given as $\boxed{\Delta G = G_2 - G_1 = RT \ln [C_2]/[C_1]}$ and $\boxed{Jc = -P\{[C_2]-[C_1]\}}$

$[C_1]$ and $[C_2]$ = Concentration of both the sides

$[C_2]-[C_1]$ = Concentration gradient

G_1 and G_2 = Potentials on both the sides

R = Gas Constant (2 cal deg^{-1} mole^{-1})

T = Absolute Temperature

P = Permeability Coefficient (cm / sec)

P = Constant (depends on how readily the water molecule leaves the water] solvent and crosses the hydrophobic membrane)

P depends on three factors namely k, D and X and it works as follows:-

$$\boxed{P = kD / X}$$

k = Partition Coefficient

k = Solubility of water in hydrophobic solvent / solubility of water molecules

D = Diffusion Coefficient

X = Thickness of the membrane

Transport of charged molecules

Transport of charged molecules depends on electrochemical potential which is stated as

$$\boxed{\text{Electrochemical Potential (or) } \Delta G = G_2 - G_1 = RT\ln\,[C_2]/[C_1] + zFV}$$

$$V = V_2 - V_1$$

z = Charge on transported molecule

F = Faraday's Constant

V = Electric Potential Difference (Voltage difference) on the two faces of the Membrane

ZFV = Accounts for the movement of a charge across a potential difference which depends on 'z' and 'V', since 'F' is a constant.

If side 2 (V_2) has higher potental than side 1 (V_1), V = Positive (ions will move from side 2 to 1) and If side 1 (V_1) has higher potental than side 2 (V_2), V = Negative (ions will move from side 1 to 2).

Passive Transport – Ion Channels

- Ion fluxes across cellular membranes mediate a variety of essential biological processes
- Involves passive processes such as **symport** (or) **antiport** involving **facilitated diffusion.**
- Facilitated diffusion is by integral membrane proteins called ion channels
- **Ion channel is a single channel protein**. These proteins have specific properties to perform specific functions.

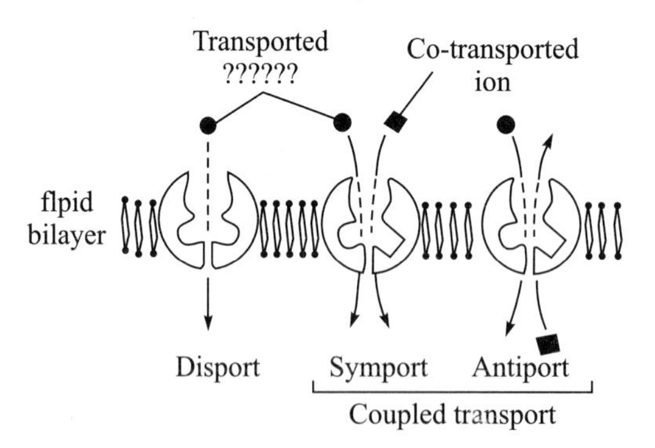

Carrier proteins of membrane tarnslicaning as disports, symports and antiports (after alberts ex al.. 1989).

Fig. 4.2 Transport by ion Channels

- Ion channels have varying pores ranging from simple aqueous to complex pores.
  a. Ion Selectivity (flow of only specific ions is permitted)
  b. Activation gating (Channels opening is regulated)
  c. Inactivation Gating (Channel closing is regulated)
    i. Channel proteins are also involved in the formation of gap junctions
    ii. It is estimated that 1 million ions can pass through 1 channel per second
    iii. 1000 times greater than the fastest rate of transport mediated by any carrier protein
    iv. Ion channels mediate only passive transport (down hill)
    v. These ion channels are used in the nerve cells for receiving, conducting and transmitting signals in the form of an 'Action Potential' (Action Potential means when ions of opposite charge are separated by a permeable membrane, there is a transmembrane electrical gradient called the action potential that is expressed in Volts or millivolts).

Types of ion channels

i. Voltage Gated Channels
ii. Mechanical Gated Channels
iii. Ligand-Gated Channels

Table 2 Families and Sub-families of Ion Channels

S. No.	*Family of Channel*	*Sub-families*	*Excitatory/Inhibitory*
1.	Voltage-gated cation channels	Na+, K+	Excitatory
2.	Mechanical gated channels	-	-
3.	Ligand-gated ion channels		
	A. Neurotransmitter-gated channels		
	(a) Cation Channels	Acetylcholine Na+ channels Serotonin gated Na+ Glutamate gated Na+	Excitatory
	(b) Anion Channels	GABA gated Cl-channels (Gamma Amino Butyric Acid) Glycine gated Cl- Channels	Inhibitory
	B. Ion - gated Channels	-	-
	C. Nucleotide gated Channels	-	-

Action potential

When a permeable membrane separates ions of opposite charge, there is a transmembrane electrical gradient called the action potential. This is expressed in Volts (or) millivolts.

The point at which there will be no net flow of ions across the plasma membrane is called **Resting Membrane Potential**.

Transporters

Membrane proteins that helps in the speeding up of the movement across the membrane are called transporters.

Table 3 Carrier Proteins Involved in Active/Passive Transport

S. No.	*Family of Transporter*	*Examples*	*Transport System*
1.	Sugar Transport	Glucose transporter	Facilitated diffusion
2.	Cation Transporting ATPases	Na+, K+-ATPases Ca2++-ATPases H+, K+-ATPases	Active Transport
3.	ABC Transporters	(i) Multidrug Resistance (MDR) Mammalain cells (ii) Periplasmic-Substarte binding Protein (PsBP) Bacteria (iii) Chloroquine-resistant ATPase (Palsmodium falciparum) (iv) Cystic Fibrosis Transmembrane Regulator (CFTR)	Active Transport
4.	Symporters	(i) Na+-Glucose Symporter (intestine) (ii) Na+-Proline Symporter (bacteria) (iii) Na+-HCO_3^- Symporter (glial cells)	Active Transport
5.	Antiporters - Cation Antiporters - Anion Antiporter - Cation / Anion Antiporter	 - Na+- H+ Exchanger - Cl^- - HCO_3^- Exchanger - Na+ dependent Cl-exchanger	Facilitated Diffusion

Passive Transport – Facilitated Transport

Example

Red Blood Corpuscles of Human Red Blood Cells. Transport of ions across the hydrophobic membrane, from a higher to a lower concentration with the help of carrier proteins is defined as facilitated transport. The carrier proteins perform the function of facilitating diffusion and these proteins share two things in common and they are:-

1. They facilitate movement of solutes in thermodynamically favored direction
2. They display affinity and specificity for the solute to be transported

As a result, facilitated diffusion reaches a saturation Point (V max) and there is an increase in the concentration of the solute. The velocity of flow is plotted against the solute concentration and the graph takes the shape of a rectangular hyperbola.

Types of carrier proteins

1. Uniporters (A single solute is transported from one side of the membrane to the other)
2. Symporters (Two solutes move in same direction across the membrane simultaneously)
3. Antiporters (Transport of one solute simultaneously with another solute in opposite direction)

Active Transport

Active transport is nothing but **transport of solutes from a lower concentration to a higher concentration**. Since movement from lower to higher is not easy, the process needs **energy**. Energy is often **derived from ATP** and rarely from **light**. For light, e.x. is **Bacteriorhodopsin protein** in the bac'teria called *Halobacterium halobium.*

Active transport is always mediated through **carrier proteins** or **pumps.** Passive transport sometimes uses this carrier proteins (very rare). **Ion channels** are never involved in active transport and they are **involved only in passive transport**.

ATP-drive active transport

ATP hydrolysis helps the following

1. Na+, K+ transport
2. Ca++ transport
3. H+, K+ transport
4. ABC transporter superfamily
5. Osteoclast proton pumps

1. Na+, K+ - ATPase system (Sodium pump) Animal cells throw out Na+ ions and take in K+ ions, coordinated by a protein called **Na+, K+ - ATPase** or sodium pump. This pump also helps in transport of sugars (**Na+-sugar**) and amino acids (**Na+-amino acid symporter**).

Inside the cell, Na+ conc. is 5-15mM and K+ is 100-140mM; but in the extracellular space the conc. is reversed. Na+ is inhibitory for K+.

Generally, Na+, K+ and Cl- and other ion gradients help neuronal communication; they regulate cell volume and shape; help transport of sugars, amino acids, nucleotides and others.

20-4-% of metabolic energy is used for the maintenance of ion gradients; 70% in neural tissues

For each ATP molecule being hydrolysed, 3 Na+ ions come out and 2 K+ ions get in. The hydrolysis of ATP by ATPase takes place in the cytoplasmic side of the membrane. Since an extra +ve charge gets out, the sodium pump is also referred as electrogenic.

ATPase exists in 2 forms: E1-form - with affinity for Na+ and ATP which helps getting out 3Na+ and E2 form – having less affinity for Na+ but more for K+ ions.

The cardiac glycosides, plant poisons, inhibit sodium pump and Na+, K+ ATPase leading to accumulation of Na+ and Ca++ in the cell causing hypertenions

2. Ca++ATPase Important in signal transduction and control of many functions including muscle contraction

Free Ca++ is lower in cytosol (10^{-7} M) relative to outside the cytosol or in ER (10^{-3}M)

All Ca++ is accumulated in **sarcoplasmic reticulums** in the cytoplasms of muscles. Excess of Ca++ is driven out by Na+ electrochemical gradient. Two Ca++ is sent to SR per ATP molecules being hydrolysed.

3. H+, K+ -ATPase The highly acidic stomach is essential for digestion. This acidity is maintained by H+ ions. The ATPase enzyme sends H+ into the stomach from mucosa in exchange of K+ ions, making the transport system electrically neutral. The Cl- ions pumped into the stomach, forming HCl.

4. ABC transporter superfamily This is for transport of peptides and drugs. These are ATPase carrier proteins. They contain ATP binding cassettes. They have clinical significance, especially in cystic fibrosis. Some of the eukaryotic ABC transporters:

a. Multi-drug resistance (MDR) protein – able to pump hydrophic drugs out of the cells through overexpression – developing resistant in cancer chemotherapy

b. Malaria resistance (*Plasmodium falciparum*) to chloroquine – through amplification of a gene encoding an ABC transporter that pumps out the drug.

c. In humans, causes cystic fibrosis, through mutation in a gene encoding an ABC transporter that functions as Cl- ion channel in the epithelial cells.

5. Osteoclast H+ pump & other H+ ATPase These are multi-nucleate cells breaking down during bone remodeling. These ATPases provide Ca for soft tissues, nerves and muscles. Ca is released by pumping of H+ in the extracellular space of bones.The protons (H+) make acids, dissolving and releasing Ca++

Others

Light-driven active transport

1. Bacteriorhodopsin or bR (H+ pump)
2. Halorhodopsin or hR (Cl- pump)

Both the above found in *Halobacterium halobium*, an archaebacterium thriving in high-salt medium. Respiration takes place in the presence of oxygen, normally. In the absence of oxygen, bR and hR use light energy through outward H+ transport.

Ion-gradient driven active transport The gradient difference of anions and cations cause **secondary active transport** of amino acids and sugars. Eg. Anion transporter of erythrocytes. Lactose is transported with each H+ ions through lactose permease.

Membrane excitability (neurotransmission and ion channels) Nerve impulses move at 100m/sec. Nervous system contains **receptor proteins** in excitable cells. Any stimulus is sent from one part to another and is reversible. The changes involve transport of ions and neurotransmitters through nerve cells called neurons.

Neurons and neuroglia (glial cells) A neuron contains: 1. cell body (nucleus and ER and mitochondria) 2. axon (cellular and myelin sheath; terminates at synaptic terminals, knobs and bulbs; gap between synaptic terminal and dendrite called synaptic cleft) and 3. dendrite

Neurons are of 3 types:

1. Sensory neurons – for sensory signals
2. Interneurons – neuron-neuron connection
3. Motor neurons – muscle movements

Ion gradients for transmission of nerve impulse or action potential Nerve impulses are called as **action potential.** It is transmitted from neuron to neuron through transient changes in the electrical potential differences (voltages) across the membranes.

At rest, neuron is rich in K+ and poor in Na+ and Cl- ions relative to extracellular fluid. This maintained by Na+, K+-ATPase. This is **resting potential** (approx. –60mV).

Action potential is created through depolarization (approx. 20mV changing to –40mV). Na+ ions enter the cell till finally the potential reaches +30mV to +50mV. At this stage, Na+ reaches equilibrium. Then Na+ ions automatically become inactive, opening up K+ channels. Then K+ also comes out. The changes in membrane potential are rapidly passed along the axonal membrane.

Neuron transmits signals to another through synapse & same neuron has several synapses (1 to 10,000).

Acetylcholine as a neurotransmitter Acetylcholine molecules (apprx. 1000 moles/vescile) are seen in the synapses. During action potential Ca++ ions enter the synapses and make Ach attach with the membrane. Ach binds to Ach receptor and cause opening of Na+ and K+ ion channels, generating new action potential.

There are muscarinic (stimulated by muscarine) and nicotinic receptors (stimulated by nicotine) of Ach found in postsynaptic membranes. Both are glycoproteins. After transmission, Ach is degraded by Ach-esterases.

There are also G-protein linked receptors and enzyme-linked receptors secreted by axon terminals. Ex. Neuropeptides.

There are few drugs having specificity for individual channels. Eg. Psychoactive drugs attack neurotransmitter-gated channels. Eg. Curare that blocks Ach receptors on skeletal muscle. Valium and Librium are tranquilizers binding to GABA receptors.

Muscle contraction involves 5 different ion channels:

1. During action potential Ca++ flows inside, which releases Ach from synaptic vesicles
2. Ach binds to Ach receptors in muscle, opening Na+ to enter cytosol causing depolarization
3. Depolarization increases more entry of Na+
4. This activates Ca++ channels within cytozol; causes sarcoplasmic reticulum to release stored Ca++ ions. This causes myofibrils in the muscle to contract.
5. The reverse occurs for muscle release.

Table 4 List of few Neurotransmitters

S. No.	*Class of Neurotransmitters*	*Neurotransmitters*
1.	Cholinergic Agents	Acetylcholine
2.	Catecholamines	Norepinephrine (Nor-adrenaline) Epinephrine (Adrenaline) L-Dopamine Octapamine
3.	Amino acids and their derivatives	GABA-Gamma Aminobutyric Acid Alanine Aspartate Glycine, Taurine
4.	Peptides	Serotonin Tyrosine Cholecystokinin Endorphins Gastrin, Gonadotropin Oxytocin Secretin Substance P Somatostatin TRF-Thyrotropin Releasing Factor Vasopressin
5.	Gases	Carbon Monoxide (CO) Nitric Oxide (NO)

Vesicular Transport & Membrane Fusion

The membrane system allows eukaryotic cells

- to take up macromolecules from cell surface through **endocytosis** by **coated pits** or **caveolae**
- to throw out substances through **exocytosis** by **vesicles**

The **endosomes** – deliver macromoles to lysosomes for digestion or throw back to cell surface for reuse. This process is called as **endocytic pathway**

Besides endocytic pathway, the eukaryotic cells deliver newly synthesized proteins and carbohydrates either to the exterior or from one compartment to another by **biosynthetic secretory pathway**. This involves exocytosis and also transport between endoplasmic reticulum and golgi and in turn to lysosomes.

The endocytic and biosynthetic pathways transport macromoles by numerous transport vesicles. They fuse with one another and transport takes place from transmembrane cytosol to various organelles and cytosol to nucleus.

Protein sorting and vesicular traffic from ER to golgi

Earlier, it was believed to take place through default pathway with no transport signals

Recently, found to be carried by vesicles called **pre-golgi intermediates** (PGI). They move on microtubules to the **cis-golgi region.** When moved through **golgi cisternae**, they are modified by **golgi-associated processing enzymes** and reach **trans-golgi** surface. Sorting takes place for various extracellular destinations.

The above is described as directed **maturation** or **cisternal progression**.

Transport of large moles (ex. procollagen precursors) is also carried out by signals without vesicles. Therefore, a protein may or may not enter a vesicular budding, the choice being decided by the sorting signals.

Also, a protein will have multiple signals depending on the fate of protein at one of the different successive steps during protein transport. There are 3 different fates for a protein by vesicular transport.

i) The protein having **transport signal** by residing on cytoplasmic side (direct binding) or luminal side (receptor binding) of ER membrane. Ex. Low density lipoprotein (LDL) receptor
ii) The protein having **retention signal** leading to lack of movement. This may escape and get transported into golgi but again retained back
iii) The protein lacks both transport and retention but having **default pathway**

Retrograde transport (Golgi to ER & vice versa)

The ER and golgi retain their own proteins, after departing vesicles. But on the way the integral membrane protein and soluble lumen protein of ER and golgi escape and then returned to the donor compartment by retrograde transport. Ex. KDEL at COOH terminal retrieves protein from golgi to ER by KDEL receptor; KKXX or XXRR help retrograde movement of protein of ER (K-lysine; R-arginine; D-aspartic acid; L-glutamic acid; X-amino acid)

Transport from CIS to trans golgi (Anterograde & Retrograde Transport)

The transport takes place from ER to cis to trans. As KDEL receptor is present in all cisternae of golgi, retrieval takes place at each step of successive cisternae. The retrograde transport takes place in 2 ways:

1. Cross current flow – direct traffic from each cisterna of golgi to ER
2. Counter current flow – percolation backwards and retained in the golgi

Sometimes, both 1 & 2, coexist in the golgi

Distillation tower hypothesis

Golgi stack of cisternae can act as distillation tower, retaining proteins meant for retrieval at any one of the stacks. Therefore, proteins get concentrated at the cis phase than trans phase and hence KDEL receptors are also seen more in cis than in trans.

Transport from Trans golgi network (TGN) to lysosomes - The received protein in golgi from ER may be **resident protein** (retained as permanent proteins) or **secretory protein** (sorted out by TGN as per the intended destination) which are destined to be transported to lysosomes, ex. Digestive enzymes.

Lysosomal hydrolases and membrane proteins for lysosomes are synthesized by rough ER. They are transported to lysosomes through golgi by means of trans-golgi network.

Mannose 6 – phosphate (M6P) selectively sorts out lysosomal protein for transport to lysosomes. M6P is recognized by M6 receptors. After delivery, they dissociate and return to trans-golgi via transport vesicles.

Transport from trans golgi network (TGN) to cell surface (Exocytosis)

Exocytosis, secretion out of the cell, starts at golgi apparatus. The vesicles leave the TGN and instead of merging with lysosomes, they arrive to cell surface and get fused with the plasma membrane. The vesicles thus carry lipids and hydrophobic membrane proteins and soluble proteins as well for regulated secretion of neurotransmitters or hormones.

There are 2 types of secretion:

1. **Constitutive secretion** – occurs in all cells – ex. Insulin, epinephrine, trypsin
2. **Regulated secretion** – occurs in specialized cells, secreting specific products – ex. Sperm acrosome, oocyte cortical granules, endocrine cells. The regulated secretion takes place rapidly but only on demand.

Generally, the secretory vesicles, once loaded reach the site of secretion and wait at plasma membrane till signal is received for secretion.

Secretoary vesicles are formed due to **clathrin** coated budding from trans-golgi network. These coats wear off during maturation.

There is an additional **pathway in polarized cells** having different delivery systems in different domains.

Secretory proteins are synthesized as inactive precursor protein molecules (pre-pro-proteins), which produce active proteins, by proteolysis. Some **polyproteins** are also synthesized for signals.

In case of nerve cells, where secretion is far away from Golgi, the secretory vesicles secrete peptide neurotransmitters at the tip of the axon, which is facilitated by motor proteins. At the axon, vesicles receive signal often-involving Ca ++ ion concentration in the form of intracellular signals. The action potential triggered by chemical transmitter causes influx of Ca ions that in turn triggers exocytosis.

Synaptic vesicles for transport of neurotransmitters

Synaptic vesicles (50nm in dia) are found in nerve cells, originated from endosomes near axon. They carry neurotranmitters such as acetylcholine, glutamate and **gamma** aminobutyric acid) to the synaptic cleft. They allow rapid signal from cell to cell through chemical synapses.

When there is action potential the synaptic vesicles secrete their contents in millisecond. The synaptic vesicles are retrieved back by endocytosis to form endosomes. From endosomes, vesicles bud off again to form synaptic vesicles. They collect neurotransmitter from cytosol. Defect in synaptic transmission leads to Parkinsonism.

Membrane Transport and Membrane Excitability

Membrane transport is of two types, namely, passive transport and active Transport

Active transport

Active transport is nothing but **transport of solutes from a lower concentration to a higher concentration.** Since movement from lower to higher is not easy, the process needs **energy**. Energy is often **derived from ATP** and rarely from **light**. For light, e.x. is **Bacteriorhodopsin protein** in the bacteria called *Halobacterium halobium.*

Active transport is always mediated through **carrier proteins** or **pumps.** Passive transport sometimes uses this carrier proteins (very rare). **Ion channels** are never involved in Active transport and they are **involved only in passive transport**.

ATP-drive active transport ATP hydrolysis helps the following

i) Na+, K+ transport
ii) Ca++ transport
iii) H+, K+ transport
iv) ABC transporter superfamily
v) Osteoclast proton pumps

1. Na+, K+ - ATPase system (Sodium pump) Animal cells throw out Na+ ions and take in K+ ions, coordinated by a protein called **Na+, K+ - ATPase** or sodium pump. This pump also helps in transport of sugars (**Na+-sugar**) and amino acids (**Na+-amino acid symporter**).

Inside the cell, Na+ conc. is 5-15mM and K+ is 100-140mM; but in the extracellular space the conc. is reversed. Na+ is inhibitory for K+.

Generally, Na+, K+ and Cl- and other ion gradients help neuronal communication; they regulate cell volume and shape; help transport of sugars, amino acids, nucleotides and others.

20-4-% of metabolic energy is used for the maintenance of ion gradients; 70% in neural tissues

For each ATP molecule being hydrolysed, 3 Na+ ions come out and 2 K+ ions get in. The hydrolysis of ATP by ATPase takes place in the cytoplasmic side of the membrane. Since an extra +ve charge gets out, the sodium pump is also referred as electrogenic.

ATPase exists in 2 forms: E1-form - with affinity for Na+ and ATP which helps getting out 3Na+ and E2 form – having less affinity for Na+ but more for K+ ions.

The cardiac glycosides, plant poisons, inhibit sodium pump and Na+, K+ ATPase leading to accumulation of Na+ and Ca++ in the cell causing hypertenions

2. Ca++ATPase Important in signal transduction and control of many functions including muscle contraction

Free Ca++ is lower in cytosol (10^{-7} M) relative to outside the cytosol or in ER (10^{-3}M)

All Ca++ is accumulated in **sarcoplasmic reticulums** in the cytoplasms of muscles. Excess of Ca++ is driven out by Na+ electrochemical gradient. Two Ca++ is sent to SR per ATP molecules being hydrolysed.

3. H+, K+ -ATPase The highly acidic stomach is essential for digestion. This acidity is maintained by H+ ions. The ATPase enzyme sends H+ into the stomach from mucosa in exchange of K+ ions, making the transport system electrically neutral. The Cl- ions pumped into the stomach, forming HCl.

4. ABC transporter superfamily This is for transport of peptides and drugs. These are ATPase carrier proteins. They contain ATP binding cassettes. They have clinical significance, especially in cystic fibrosis. Some of the eukaryotic ABC transporters:

a. Multi-drug resistance (MDR) protein – able to pump hydrophic drugs out of the cells through overexpression – developing resistant in cancer chemotherapy
b. Malaria resistance (*Plasmodium falciparum*) to chloroquine – through amplification of a gene encoding an ABC transporter that pumps out the drug.
c. In humans, causes cystic fibrosis, through mutation in a gene encoding an ABC transporter that functions as Cl- ion channel in the epithelial cells.

5. Osteoclast H+ pump & other H+ ATPase These are multi-nucleate cells breaking down during bone remodeling. These ATPases provide Ca for soft tissues, nerves and muscles. Ca is released by pumping of H+ in the extracellular space of bones. The protons (H+) make acids, dissolving and releasing Ca++

Others

Light-driven active transport

1. Bacteriorhodopsin or bR (H+ pump)
2. Halorhodopsin or hR (Cl- pump)

Both the above are found in *Halobacterium halobium*, an archaebacterium thriving in high-salt medium. Respiration takes place in the presence of oxygen, normally. In the absence of oxygen, bR and hR use light energy through outward H+ transport.

Ion-gradient driven active transport The gradient difference of anions and cations cause **secondary active transport** of amino acids and sugars. For example, Anion transporter of erythrocytes. Lactose is transported with each H+ ions through lactose permease.

Membrane excitability *(neurotransmission and ion channels)* Nerve impulses move at 100m/sec. Nervous system contains **receptor proteins** in excitable cells. Any stimulus is sent from one part to another and is reversible. The changes involve transport of ions and neurotransmitters through nerve cells called neurons.

Neurons and neuroglia *(glial cells)* A neuron contains: 1. cell body (nucleus and ER and mitochondria) 2. axon (cellular and myelin sheath; terminates at synaptic terminals, knobs and bulbs; gap between synaptic terminal and dendrite called synaptic cleft) and 3. dendrite

Neurons are of 3 types:

1. Sensory neurons – for sensory signals
2. Interneurons – neuron-neuron connection
3. Motor neurons – muscle movements

Ion gradients for transmission of nerve impulse or action potential Nerve impulses are called as **action potential.** It is transmitted from neuron to neuron through transient changes in the electrical potential differences (voltages) across the membranes.

At rest, neuron is rich in K+ and poor in Na+ and Cl- ions relative to extracellular fluid. This maintained by Na+, K+-ATPase. This is **resting potential** (approx. –60mV).

Action potential is created through depolarization (approx. 20mV changing to –40mV). Na+ ions enter the cell till finally the potential reaches +30mV to +50mV. At this stage, Na+ reaches equilibrium. Then Na+ ions automatically become inactive, opening up K+ channels. Then K+ also comes out. The changes in membrane potential are rapidly passed along the axonal membrane.

Neuron transmits signals to another through synapse and same neuron has several synapses (1 to 10,000).

Acetylcholine as a neurotransmitter Acetylcholine molecules (apprx. 1000 moles/vescile) are seen in the synapses. During action potential Ca++ ions enter the synapses and make Ach attach with the membrane. Ach binds to Ach receptor and cause opening of Na+ and K+ ion channels, generating new action potential.

4.3 PERMEASES

Permeases are transport proteins (or) carrier proteins present in the plasma membrane of bacteria. Permeases, by its ending 'ase' represent an enzyme too and are associated more with the function of proteins. Kennedy isolated and labeled a protein from the plasma membrane of a bacteria called E.coli and named it as permeases.

Function

a) Accelerates the reaction process in bacteria
b) Membrane translocation
c) Provide great selectivity
d) Remains unchanged in the process and are recycled after each entry and exit

Example

- E.coli (bacteria)
- Inner mitochondrial membrane also contains several permease systems

Characteristic Features

a) The plasma membrane of E.coli contains a specific chemical transport system called permease system
b) Permease is the enzyme that determines the entry of lactose into the bacterium
c) β-galactosidase also enters into the bacterium through the enzyme, permease
d) This permease is considered primarily as a protein and are therefore, egentically determined and includes a specific membrane protein specialized for transport
e) This permease protein in E.coli increases the rate of reaction till an equilibrium is reached

4.4 PUMPS IN THE TRANSPORT SYSTEM

Carrier proteins bind specific solutes and transfer them across the lipid bilayer by undergoing conformational changes that expose the solute-binding site sequentially on one side of the membrane and then on the other. Some carrier proteins simply transport a single solute "down hill" whereas others can act as pumps to transport a solute "up hill" against its electrochemical gradient, using energy provided by ATP hydrolysis or by a "down hill" flow of another solute (such as Na^{+}) to drive *the requisite series of conformational changes. DNA cloning and sequencing studies show that carrier proteins belong to a small number of families, each of which comprises proteins of similar amino acid sequence that are thought to have evolved from a common ancestral protein and to operate by a similar mechanism. The family of Cation-transporting ATPases, which includes the ubiquitous Na^+-K^+ pump, is an important example. Each of these ATPases contains a large catalytic subunit that is sequentially phosphorylated and dephosphorylated during the pumping cycle. The super family of ABC transporters is especially important clinically. It includes proteins that are responsible for cystic fibrosis as well as for drug resistance in cancer cells and in malaria causing parasites.*

Sodium – Potassium Pumps (Na^+, K^+- ATPase Pump)

The Sodium – Potassium Pump was discovered by A.L.Hodkin and R.D. Keynes in the year 1955.

The Na^+, K^+ - ATPase System commonly called as sodium pump, is an ATP dependent active transport system. In this sodium pump, outward pumping of Na+ is tightly linked to the inward transport of K+ ions. Since Na+ and K+ are exchanged in a compulsory way, inward movement of K+ always accompanies an outward movement of Na+.

Example

Sodium pump, commonly occurs in nerve and muscle cells

Characteristic Features

a) All animal cells actively throw out Na+ ions and take in K+ ions, and an integral protein called Na+, K+ - ATPase, facilitates these processes.

b) Sodium pump works in an antiporter mechanism

c) In all cells, Na+ concentration is lower within the cell (5-15mM) and K+ concentration is higher (100 – 140mM) within the cell and vice-versa, outside the cell or Extra cellular membrane.

d) The energy required to pump Na+ ions out of the cell is provided by ATP, which is hydrolysed by an enzyme called ATPase. Sometimes, hydrolysis is believed to be facilitated by a Mg+ activated ATPase believed to be situated within the membrane

e) K+ is needed for many functions such as

   i. To help the neurons to communicate

   ii. To regulate the cell volume and cell shape

   iii. To help in transport of amino acids, sugars, nucleotides and other substances

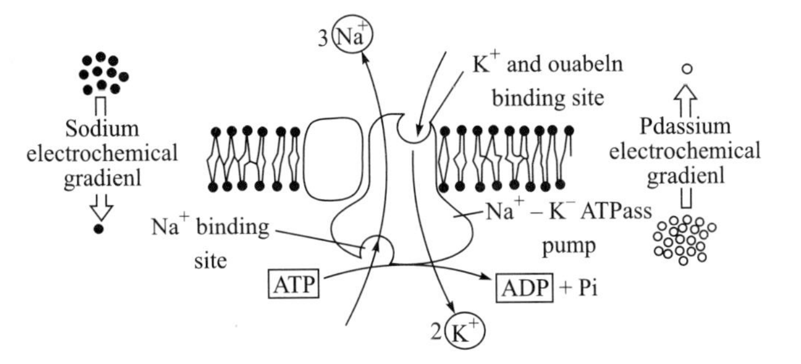

Fig. 4.3 The Na+-K+ ATPase Pump (After Alberts *et al.*, 1989)

f) ATPase molecule consist of

i. a subunit (120kD) consisting of 10 transmembrane helices and two large cytoplasmic domains
ii. a β subunit (35kD) consisting of 1 transmembrane segment and a large extracellular domain

g) The ATPase hydrolyses ATP on the cytoplasmic side of the membrane so that three Na+ ions are transported out of the cell and two K+ ions are transported into the cell for each ATP molecule hydrolysis

h) Since this involves net movement of one positive charge outward per cycle, the sodium pump is described as '**electrogenic**' in nature

i) The ATPase exists in two forms

a. **E_1 form** : has affinity for Na^+ and ATP. E_1 form gets readily phosphorylated in the presence of ATPase or Mg^{++} to form E_1-P carrying three Na^+ ions.

b. **E_2 form** : has low affinity for Na+ ions but high affinity for K+ ions. E1 – P changes into E2 – Pi and this E2 – Pi releases 3 Na+ ions and binds 2K+ ionson the outside of the cell giving rise to E2-2K+ releasing 2K+ to the interior. The net process is written as follows:-

ATP + H2O+3Na+ (inside) + 2K+ (outside)→ ADP+Pi+H2O+3Na+(outside)+ 2K+ (Inside)

j) Any inefficient working of this sodium pump results in swelling or bursting of cells and narrowing of blood vessels (hypertension)

Calcium – ATPase Pumps or Calcium Pump

a) The Ca2+-ATPase carrier protein is present in the plasma membrane and in several organelle membranes including the membranes of the endoplasmic reticulum and the mitochondria.
b) In the plasma membrane, the direction of Ca++-transport is from intracellular to extracellular fluid
c) This results in the cytosol of most cells have a very low Ca++ concentration compared with the extracellular Ca+ that is 10,000 times greater
d) The cytosolic Ca+ (calcium) will be increased through the help of calcium channels. This Ca++ channels allow ions to diffuse from high concentration to low, into the cytosol.

Function

- An increase in the cytosolic Ca++ ions leads to chemical signalling or signal transduction pathways.
- Muscle contraction and relaxation (sarcoplasmic reticulum, SR)
- Ca++ pump works in an antiporter mechanism
- An enzyme called Ca++-ATPase is found, similar to Na+, K+-ATP ase enzyme. Their mechanism and function are also similar in many ways.

Hydrogen Pump or H^+, K^+- ATPase

a) The hydrogen – potassium Pump carrier is present in the plasma membrane and in several organelle membrane including inner mitochondrial membrane.
b) In the plasma membrane, the Hydrogen pumps transports hydrogen ions out of the celkls and in exchange gets back potassium ions making the transport electrically neutral.
c) Potassium (K+) ions are pumped back out of the cells together with Cl- ions, thus making HCl inside the stomach
d) Inside the stomach, a highly acidic medium is maintained to facilitate digection of food

e) H^+, K^+- ATPase enzyme acts as an enzyme in this hydrogen pump
f) This K+ and Cl- ions being pumped back out of the cells together are also called as Co Transport
g) Osteoclast Hydrogen pump also has the same property as that of Hydrogen Pump.

Function Oxidative phosphorylation

Lysosomal and Vacular Membrane ATP Dependent Proton Pumps

a) Proton pumps (or) H+-ATPases are present in vacuoles, lysosomes, endosomes Colgi complex etc. They also occur in yeast and bacteria and perform varied functions. Smaller (or) lower organisms, when depend on energy for transportation from light and since this energy pumping action is present in the vacuolar and lysosomal regions, it is called lysosomal and vacuolar membrane ATP dependent proton pumps.
b) Basically, energy for the lower organisms are acquired from light and are therefore classified into two classes namely
   A. Bacteriorhodopsin or bR (light driven H^+ pump)
   B. Halorhodopsin or hR (Cl^- Pump)
c) Both these light pigments are present in the bacteria, *Halobacterium halobium*, an arcahebacterium, surviving in high salt media, NaCl.
d) This *Halobacterium halobium* respires normally, if O_2 is plentiful
e) In the absence (or) shortage of O_2, bR and hR capture light energy, through transport of protons
f) Absorption of a photon by bR_{568} form of bR converts all trans-configuration to the 13-cis isomer
g) Passage through several intermediate steps result in the outward export of 2H+ ions per photon absorbed by bR
h) This establishes a proton gradient, later used to the ATP synthesis and movement of molecules across the membrane
i) In addition to the outward transport of H^+ ions facilitated by bR_{568}, inward transport of Cl^- ions are also mediated by hR.
j) These light (ion gradient) driven active transport system established by ATPase or light driven transport leads to secondary active transport
k) Secondary active transport leads to the movement of substances like amino acids and sugars
l) These substances may move along with ions in the same direction (symport) or they may move along with ions in the opposite direction (antiport)

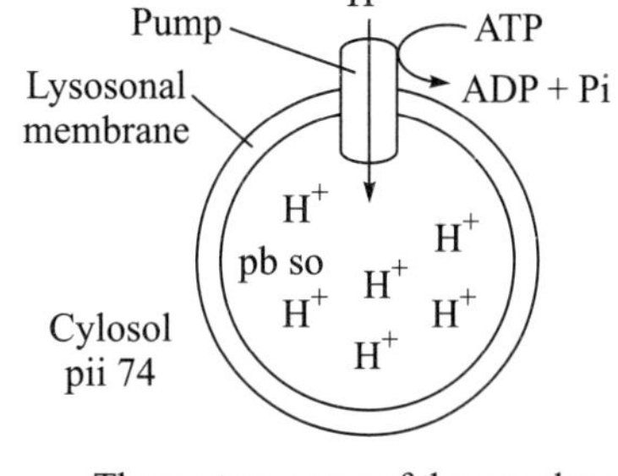

The proton pump of the membrane of the lysosome (after darrell elal., 1986).

Fig. 4.4 Proton Pump (After Darnell *et al.*, 1986)

Example

- Antiport proton pump : Transport of erythrocytes
- Symport Proton pump : Transport of lactose using lactose permease enzyme inside E.coli (LacY)

LacY

One molecule of Lactose is transported with each H+ ion establishing a proton (or) electrical gradient Lactose permease (LacY) is an intergral membrane protein

Co-transport, Symport and Antiport Systems

Co-transport system is a transport of any two solutes simultaneously across the membrane

Example

Na^+,K^+ - ATPase Pump

H^+,K^+ - ATPase Pump

Co-transport is indefinite of the directions, the solute travels and according to the mode of travel direction, it is again classified into two types such as **symport** and **antiport.**

Symport

Two solutes or subatances moving simultaneously in the same direction

Antiport

Two solutes (or) substrates moving simultaneously in the opposite direction

Illustration

- *Chloride – Bicarbonate Exchanger*

Carbon dioxide is released and Oxygen enters the lungs simultaneously during the respiratory process

- *Glucose transporter*

Cl-, HCO3- exchanger is also called anion exchange proteinand this mediates simultaneous movement of Cl- ions in one direction and HCO3- ions in the opposite direction. If movement of Cl- ions stops, HCO3- movement is also inhibited. Therefore, glucose transport needs both ions to travel simultaneously always and is called as Co-transport.

Questions

1. What are Intrinsic and Extrinsic Proteins?
2. What are the four major classes of lipids in the plasma membrane?
3. What is the chemical potential difference in passive transport?
4. What re the types of ion channels?
5. Define Action Potential and Resting Membrane Potential:-
6. What are ABC transporters?
7. Define Symporters and Antiporters:-
8. What are the types of carrier proteins?
9. What is Bacteriorhodopsin?
10. What are Osteoclast Proton Pumps?

11. What is Halorhodopsin?
12. Define Membrane Excitability:-
13. What are neurons and neuroglia?
14. What is Acetylcholine's role in a neurotransmitter?
15. What are GABA receptors?
16. What is Cisternal progression?
17. Define Retrograde transport and Anterograde transport:-
18. Explain the Distillation Tower hypothesis:-
19. What are the two types of secretion?
20. What are the different types of pumps?
21. Define Co-transport, symport and aniport systems:-
22. Explain LacY in detail:-
23. What is a glucose transporter?
24. Explain in detail the structure of plasma membrane:-
25. What are Permeases? Explain in detail:-

CHAPTER 5

Receptors and Models of Extra-cellular Signalling

5.1 RECEPTORS

Cells communicate by cell signalling. The importance of cell signalling is that in prokaryotes, there is no cell signalling since they are unicellular. In eukaryotes, cell signalling takes place at both intracellular and extracellular level. Cell signalling takes place between cells for growth and metabolism to take place and this is mediated by the presence of (or) secretion of extra-cellular signalling molecules. These extra-cellular signalling molecules are nothing but primarily **hormones**. Hormones are secreted by the autocrine / paracrine / endocrine organs in a body. These extra-cellular signalling molecules execute the functions in a cell (or) bring response in a cell (or) really help to communicate through **receptors**. Receptors facilitate the cell to respond.

Receptors are **proteins** (or) **glycoproteins**. Receptors specifically bind to **ligands**. Ligands are anything that is mobile and would bind to a receptor. After binding, it produces the destined biological action in the cell. Example – **hormones**. Insulin is a hormone that binds to an insulin receptor. Ligands are only hormones.

Hormones involved in cell signalling are protein, peptide, steroids (gluco-corticoids, estrogens, progesterone, testosterone etc), vitamin D, thyronine (T3, T4).

Cells in a multicellular organism must communicate with one another in order to direct and regulate growth, development and organization. Animal cells communicate by secreting chemicals that signal to distant cells, display cell surface chemicals that influence other cells in direct physical contact, and communicate directly via porous cellular junctions called gap junctions.

Endocrine signalling occurs when substances (hormones) are secreted by cells and travel in the bloodstream to distant target cells. In paracrine signalling, cells secrete local chemical mediators that act only on cells in the immediate environment. Paracrine signalling molecules are rapidly internalized, destroyed or immobilized such that their effects can be limited to the local environment. Synaptic signalling occurs when molecules are released vesicles at specialized neuronal cell junctions called synapses. The molecules, neurotransmitters, diffuse across the synaptic cleft and act only on the postsynaptic target cell. Protein receptor molecules on or within the target cells bind to the hormone, paracrine or neurotransmitter and a response is initiated. Often the same molecules are endocrine, paracrine or neurotransmitter, the differences lie in the rapidity and selectivity of the delivered signal.

The receptor mechanisms vary for cellular communication molecules based on their solubility in water. Those, such as neurotransmitters and proteins are water soluble and cannot cross the cell membrane without help. Others such as lipid soluble steroids can cross the lipid bilayer to bind to intracellular receptors. These hydrophobic molecules must be carried in the blood stream bound to transport proteins and therefore their half-life in the bloodstream is hours to days in contrast to hydrophillic molecules which are broken down within seconds. Therefore, water soluble signalling molecules usually mediate responses of short duration, while hydrophobic molecules mediate longer lasting responses.

Intracellular Receptors

Small hydrophobic signalling molecules (steroid and thyroid hormones) pass through the target cell membrane to bind to intracellular receptors in the cytoplasm or nucleus. The hormone receptor complex undergoes a conformational change that increases the receptors affinity for DNA and enables it to bind to specific genes in the nucleus and regulate transcription. Binding to specific genes activates or suppresses transcription of those genes. DNA recognition sites associated with steroid-hormone-responsive genes function as receptor dependent transcriptional enhancers. The products of some of these genes may in turn activate other genes to produce a delayed secondary effect.

In a cell, these receptors are found either in

A. The plasma membrane- (cell surface receptors)
B. Cytoplasm
C. Nucleus

5.2 CELL SURFACE RECEPTORS (OR) MEMBRANE BOUND RECEPTORS

The receptors that are found on the plasma membrane are called 'cell-surface' receptors.

Example

Extra-cellular signalling molecules (hormones), Proteins and peptides.

a. All water soluble signalling molecules (including neurotransmitters, protein hormones and protein growth factors) as well as some lipid soluble ones bind to specific receptor proteins on the surface of the target cells, they influence.

Intra-cellular Receptors

The receptors that are found in the cytoplasm and nucleus are called intra-cellular receptors.

Example

Steroids, Vitamin D, Thyronine (in the cytoplasm and nucleus)

In all tissues, the genetic constitution is the same (or) the gene that is present in the tissues are same BUT they perform different functions depending on its location. It is not that one cell will bind to only one ligand (or) hormone BUT different cell can bind to the same hormone (or) ligand. Depending on the sort of receptors, the cell has got and whether the receptors are present in the plasma membrane, cytoplasm or nucleus, the nature of the reaction that has to be initiated in a cell by the receptor. Example – If insulin has to be produced, then insulin genes will be expressed. Example – Acetylcholine (Ach)

Acetylcholine is a ligand / hormone. Acetylcholine binds to the receptors in skeletal muscles and also in cardiac muscles. In skeletal muscles, Acetylcholine stimulates contraction. In cardiac muscles, acetylcholine decreases the rate of contraction.

A receptor should have two important sites of function namely

A. Recognition Domain (which ligand it is?)
B. A domain that relays the information inherent in 'Ligand-Receptor complex' to various intra-cellular sites in a cell.

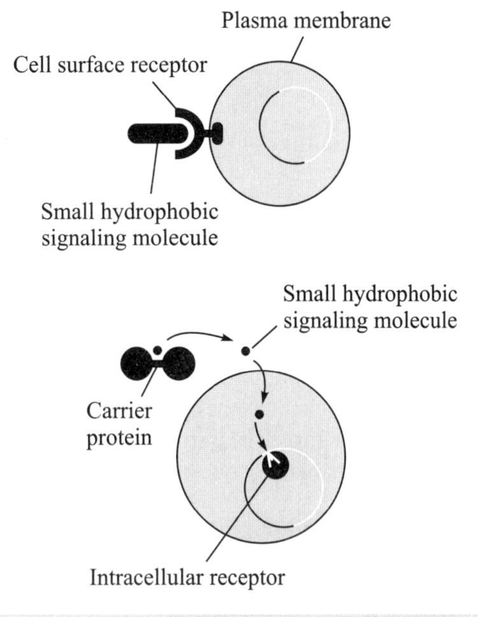

Fig. 5.1 Extra Cellular Signalling Molecules – bind either to the Cell Surface Receptors or to Intracellular Receptors

Effectors

a. These cell surface receptors act as signal transducers
b. These cell surface receptors bind the signalling ligand with high affinity and convert this extra-cellular event into one or more intra-cellular signals that alter the behavior of the target cell.
c. All cell surface receptors belong to one of these three classes defined by the transduction mechanism
d. The **three classes of transduction mechanism** are
   (1) Ion- channel linked receptors (transmitter-gated ion channels)
   (2) G-Protein linked receptors
   (3) Enzyme-linked receptors

(1) Ion- channel linked receptors (transmitter-gated ion channels)

A transmitted gated ion channel is mediated by a small number of neurotransmitters. These neurotransmitters transiently open (or) close the ion channel formed by a protein. These are involved in rapid synaptic signalling between electrically excitable cells. A neuron transmits a signal to another neuron through a synapse and the same neuron may involve in several synapses (1 to 10000). In mid-brain, there is only one synapse per post synaptic cell. The ratio of synapses to neuron is 40000:1.

Synapses may be electrical synapses or chemical synapses. In electrical synapsis, the gap between pre-synaptic and post-synaptic cells is small, only 0.2 nm. In chemical synapsis, this gap is 20 nm – 50 nm wide. In electrical synapsis, the pre-synaptic membrane leads directly to action potential whereas in chemical synapsis, the pre-synaptic membrane causes a chemical secretion called the neurotransmitter.

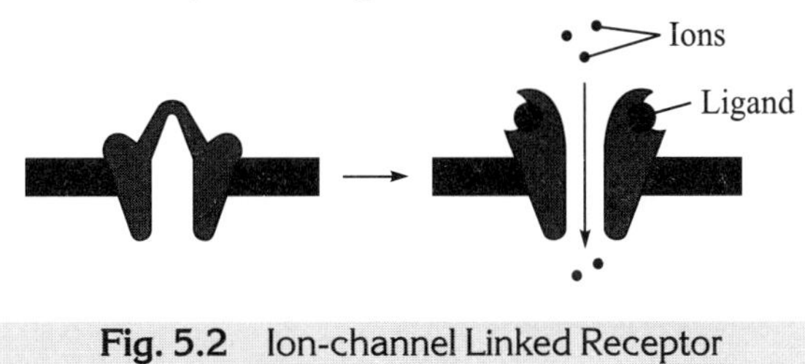

Fig. 5.2 Ion-channel Linked Receptor

(2) G-Protein linked receptors

These act indirectly to regulate the activity of a separate plasma membrane bound target protein, which can be an enzyme (or) an ion-channel. The interaction between a receptor

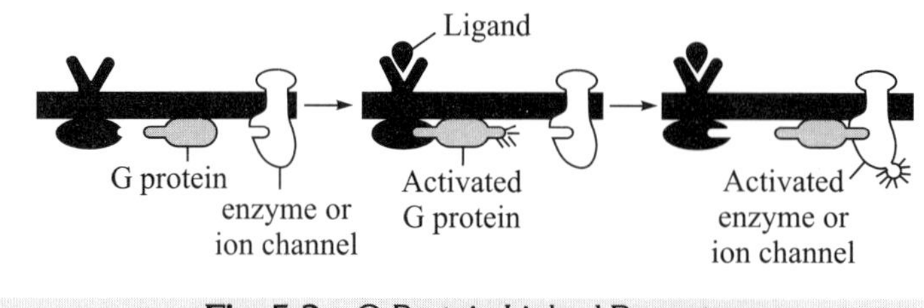

Fig. 5.3 G-Protein Linked Receptor

and the target protein is mediated by a third protein called a trimeric GTP-binding regulatory protein called G protein.

G protein is nothing but it is a trimeric (α, β, γ) GTP binding regulatory protein. The activation of the target protein either alters the concentration of the intracellular mediators (or) alter the ion-permeability of the plasma membrane. The intracellular mediators in turn alter the behavior of other proteins in the cell. All of the G proteins belong to a super-family of seven pass transmembrane proteins

(3) Enzyme-linked receptors

Enzyme linked receptors are single pass-trans-membrane proteins with their ligand binding site outside the cell and their catalytic site inside their cell. The enzyme-linked receptors are heterogenous and when activated, they function directly as enzymes (or) indirectly associated with enzymes. Majority are protein kinases (or) associates of protein kinases.

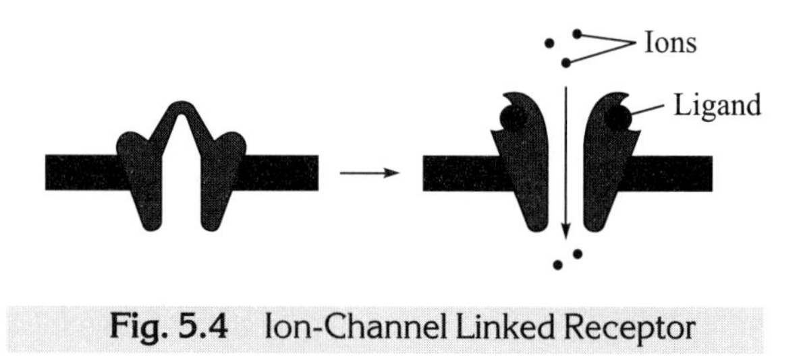

Fig. 5.4 Ion-Channel Linked Receptor

The information inherent in 'ligand-receptor complex' is relayed to various intra-cellular parts of the cell by means of effectors. Example – Adenylate cyclase and Guanylate cyclase

Function of Effector

Effector functions by means of generating a **'second messenger'.**

Examples

A. Cyclic AMP (cAMP)
B. Cyclic GMP (cGMP)
C. Protein Kinases
D. Calcium ions
E. Inositol triphosphate IP_3
F. Diacylglycerol (DAG)

Function of Second Messenger

Second messenger mimics the action of hormones (extra-cellular signalling molecules) inside a cell and so they are called the second messenger (intra-cellular).

First Messenger

All extra-cellular signalling molecules (hormones) that is shuttling between the cells and is involved in inter-cellular communication and are called the first messenger.

Signal Transduction

Transfer of signals across the membrane is called as signal transduction. The ligand binds to a receptor and the receptor is located on the surface of the plasma membrane.

Process of Signal Transduction

(a) When the ligand becomes attached to the receptor either it becomes activated or inactivated.
(b) During both these processes, the effectors recognize the message encoded in the 'ligand-receptor complex' and relays to various sites inside the cell by means of generating the second messenger.
(c) These collective net-work is called signal transduction

Autocrine

The ligand gets attached to the receptor that is present in the same cell.

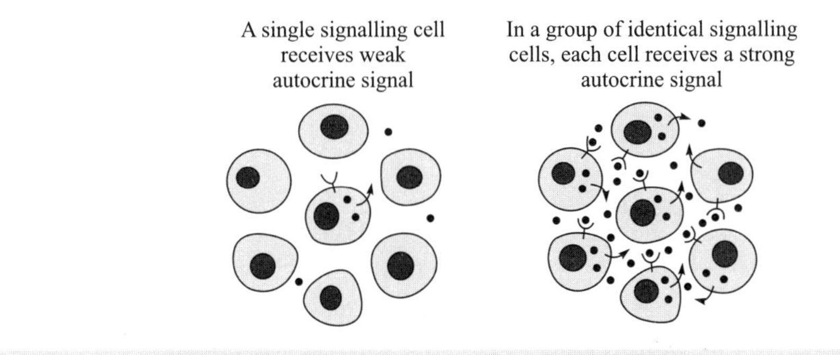

Fig. 5.5 Autocrine Signalling (after Alberts *et al.*, 1989)

Paracrine

The ligand gets attached to the receptor that is present in the adjacent cell.

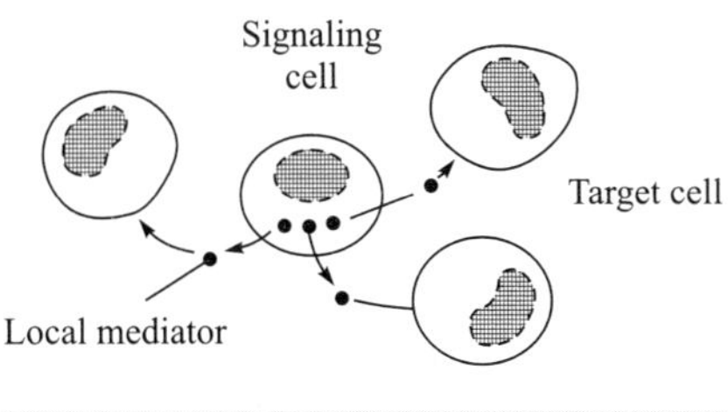

Fig. 5.6 Paracrine Signalling

Endocrine

The ligands of the signalling cells move through the blood stream and get attached to the receptors of target cells.

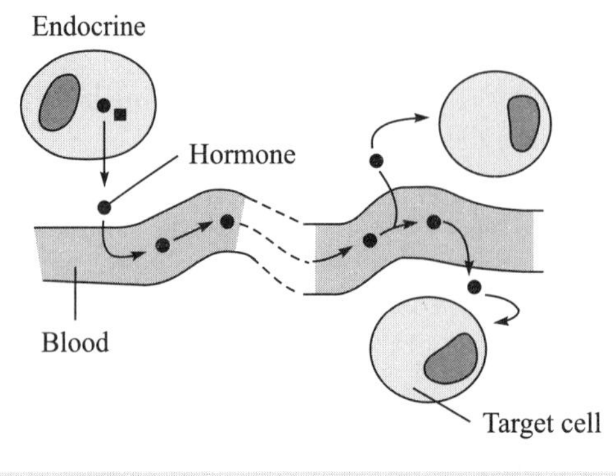

Fig. 5.7 Endocrine Signalling

Signalling cell

The cell that secretes the signal or ligand is called signalling cell

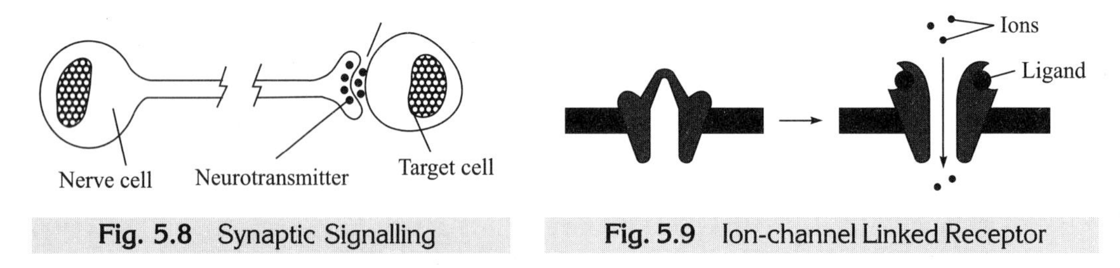

Fig. 5.8 Synaptic Signalling

Fig. 5.9 Ion-channel Linked Receptor

(1) Ion Channel-linked Receptors

These are transmitter gated ion channels involved in rapid synaptic signalling as in nervous tissue or the neuromuscular junction. A specific transmitter can rapidly open or close ion channels upon binding to its receptor thus changing the ion permeability of the cell membrane. All of these receptors belong to a family of similar multipass transmembrane proteins.

(2) G-Protein Linked Receptors

These act indirectly to regulate the activity of a separate plasma membrane bound target protein, which can be an enzyme (or) an ion-channel. The interaction between a receptor and the target protein is mediated by a third protein called a trimeric GTP-binding regulatory protein called G protein.

When bound to a specific ligand these receptors indirectly activate or inactivate a separate plasma membrane bound enzyme or ion channel. The interaction between the receptor and the affected enzyme or ion channel is mediated by a GTP binding protein. G-protein linked receptors initiate a cascade of chemical events within the target cell that usually alter the concentration of small intracellular messengers such as cAMP or inositol triphosphate. These intracellular messengers in turn alter the the behaviour of other intracellular proteins. cAMP levels affect cells by stimulating cyclic AMP-dependnent protein kinase to phosphorylate specific target proteins. Calcium levels modify the activity of certain enzymes by binding to the calcium binding protein calmodulin that then activates target proteins. The effects of all these second messengers are rapidly reversible when the extracellular signal is removed. The response of cells to external signals initiates signalling cascades that can greatly amplify and regulate various inputs.

Signalling molecules may trigger:

1. An immediate change in the metabolism of the cell (e.g., increased glycogenolysis when a liver cell detects adrenaline);
2. An immediate change in the electrical charge across the plasma membrane (e.g., the source of action potentials);
3. A change in the gene expression - transcription - within the nucleus. (These responses take more time.)

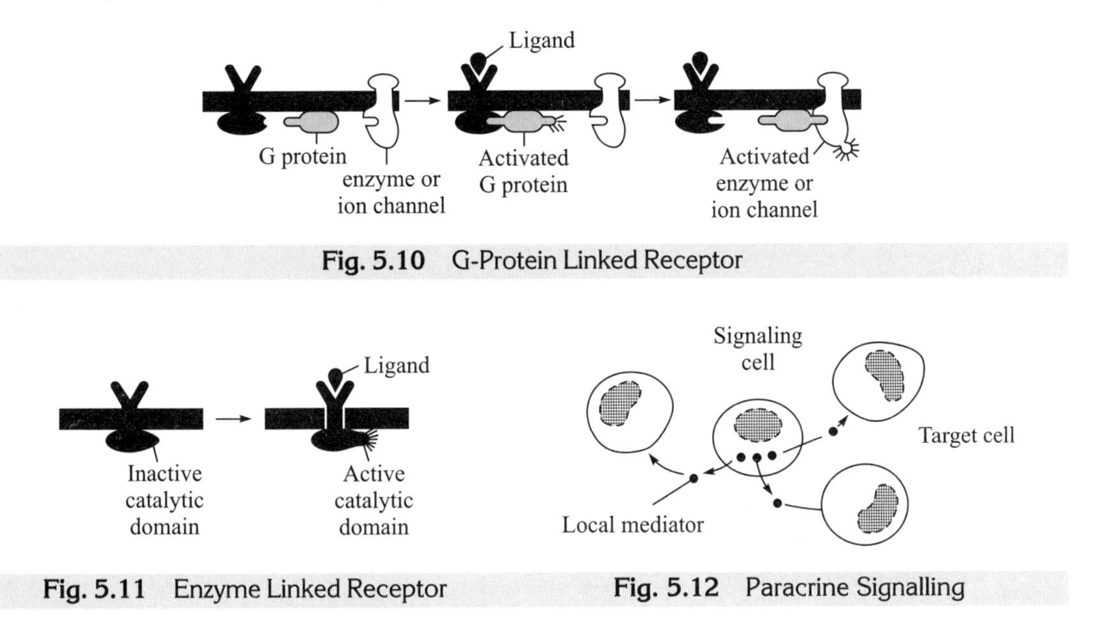

Fig. 5.10 G-Protein Linked Receptor

Fig. 5.11 Enzyme Linked Receptor

Fig. 5.12 Paracrine Signalling

(4) Catalytic Receptors

These receptors behave as enzymes when activated by a specific ligand. Most of these have a cytoplasmic catalytic region that behaves as a tyrosine kinase. Target proteins are phosporylated at specific tyrosine residues thus changing their activation state. The insulin receptor functions in this way.

(5) Steroid Receptors

Steroids are small hydrophobic molecules that can freely diffuse across the plasma membrane, through the cytosol, and into the nucleus.

Ligands - Some steroids that regulate gene expression:

- glucocorticoids (e.g., cortisol)
- mineralocorticoids (e.g., aldosterone)
- sex hormones such as
  - o estradiol
  - o progesterone
  - o testosterone
- ecdysone

Mechanism

- The steroid binds its receptor.
- The complex
  - o releases the HDACs and recruits histone acetylases (HATs) relieving chromosome repression
  - o binds to a specific DNA sequence - the Steroid Response Element (SRE) - in the promoters of genes it will turn on.

(6) Nitric Oxide (NO) Receptors

- NO diffuses freely across cell membranes.
- There are so many other molecules with which it can interact, that it is quickly consumed close to where it is synthesized.
- Thus NO acts in a paracrine or even autocrine fashion - affecting only cells near its point of synthesis.

The signalling functions of NO begin with its binding to protein receptors in the cell. The binding sites can be either:

- a metal ion in the protein or
- one of its S atoms (e.g., on cysteine).

Mechanisms

In either case, binding triggers an allosteric change in the protein which, in turn, triggers the formation of a "second messenger" within the cell. The most common protein target for NO seems to be guanylyl cyclase, the enzyme that generates the second messenger cyclic GMP (cGMP)

(7) G-Protein-Coupled Receptors (GPCRs)

These are transmembrane proteins that wind 7 times back and forth through the plasma membrane.

- Their ligand-binding site is exposed outside the surface of the cell.
- Their effector site extends into the cytosol.

Some of the many ligands that alter gene expression by binding GPCRs:

- protein and peptide hormones such as:
  - o thyroid-stimulating hormone (TSH)
  - o ACTH
- some mammalian pheromones
- leukotrienes
- GABA (which affects gene expression in addition to its role as a neurotransmitter)

Mechanisms

- The ligand binds to a site on the extracellular portion of the receptor.
- Binding of the ligand to the receptor
  - o Activates a G protein associated with the cytoplasmic C-terminal.
  - o This initiates the production of a "second messenger". The most common of these are
    - cyclic AMP, (cAMP)which is produced by adenylyl cyclase from ATP
    - inositol 1,4,5-trisphosphate (IP_3)

- o The second messenger, in turn, initiates a series of intracellular events (shown here as short arrows) such as
  - phosphorylation and activation of enzymes
  - release of Ca^{2+} stores within the cytoplasm
- o In the case of cAMP, these enzymatic changes activate the transcription factor CREB (cAMP response element binding protein)
- o Bound to its response element
  5' TGACGTCA 3'
  in the promoters of genes that are able to respond to the ligand, activated CREB turns on gene transcription.
- o The cell begins to produce the appropriate gene products in response to the signal it had received at its surface.

(8) Cytokine Receptors

Dozens of cytokine receptors have been discovered. Most of these fall into one or the other of two major families:

1. Receptor Tyrosine Kinases (RTKs) and
2. Receptors that trigger a JAK-STAT pathway.

(9) Receptor Tyrosine Kinases (RTKs)

The receptors are transmembrane proteins that span the plasma membrane just once.

Ligands

Some ligands that trigger RTKs:

- Insulin
- Vascular Endothelial Growth Factor (**VEGF**)
- Platelet-Derived Growth Factor (**PDGF**)
- Epidermal Growth Factor (**EGF**)
- Macrophage Colony-Stimulating Factor (**M-CSF**)

Mechanisms

- Binding of the ligand to two adjacent receptors forms an active dimer.
- This activated dimer is a **tyrosine kinase**; an enzyme that attaches phosphate groups to certain tyrosine (Tyr) residues - first on itself, then on other proteins converting them into an active state.
- Many of these other proteins are also tyrosine kinases and in this way a cascade of expanding phosphorylations occurs within the cytosol.

Transforming Growth Factor-beta (TGF-β) Receptors: SMADs

These are single-pass transmembrane proteins that, when they bind their ligand, become kinases that attach phosphate groups to **serine** and/or **threonine** residues of their target proteins.

Ligands for these receptors include:

- Transforming Growth Factor-beta (hence the name)

- activins
- Bone Morphogenic Proteins (**BMPs**)

Mechanisms

- The ligand binds to the extracellular portion of the receptor,
- which then phosphorylates and forms a dimer with a second transmembrane receptor.
- The dimer phosphorylates a **SMAD** protein in the cytosol.

The T-cell Receptor for Antigen (TCR)

T-cells use a transmembrane dimeric protein as a receptor for a particular combination of antigen fragment nestled in the cleft of a glycoprotein encoded by genes in the major histocompatibility complex.

5.3 CHARACTERISTICS OF RECEPTORS

1. Each Cell Influences the Behavior of Other Cells in Unicellular Organisms (Intra-cellular Signalling)

Example – Yeast (Saccharomyces cerevisiae)

The process in independent and during reproduction, budding, the haploid individual that is ready to mate secretes a peptide mating factor that signals cells of opposite mating types to stop proliferating and prepare for conjugation. Intracellular signalling in animals includes signalling by secreted molecules and signalling by plasma membrane bound molecules.

2. Extra-cellular Signalling Molecules are Recognized by Specific Receptors on (or) in the Target Cells

Extracellular signalling molecules bind either to the cell surface receptors (or) intracellular receptors.

Cell surface receptors – example : yeast. In yeast, the extracellular signalling molecules, are hydrophilic and therefore are unable to cross the plasma membrane. Therefore, they bind to the cell-surface receptors which in turn generates one or more signals inside the target cells.

Intracellular receptors – example : Higher eukaryotes. The extra-cellular signalling molecules are hydrophobic and they diffuse across the plasma membrane and bind to the receptors of the target cells cytoplasm or nucleus. The extra-cellular signalling molecules, sometimes, get the help of carrier proteins to traverse across the membrane.

3. The Secreted Molecules Mediate Four Forms of Signalling Including the Synaptic Signalling

(a) Autocrine The signalling and target cell are the same and the ligands get binded to the receptor in the same cell.

(b) Paracrine The ligand of the signalling cells bind to the target cells on the immediate environment. These cannot diffuse too far and sometimes, gets the help of local mediators

(c) Endocrine Endocrine cells secrete the signalling molecules called hormones and they are carried by the blood stream in animals and sap in plants. These signalling cells distributes signals to target cells that are distributed throughout the body.

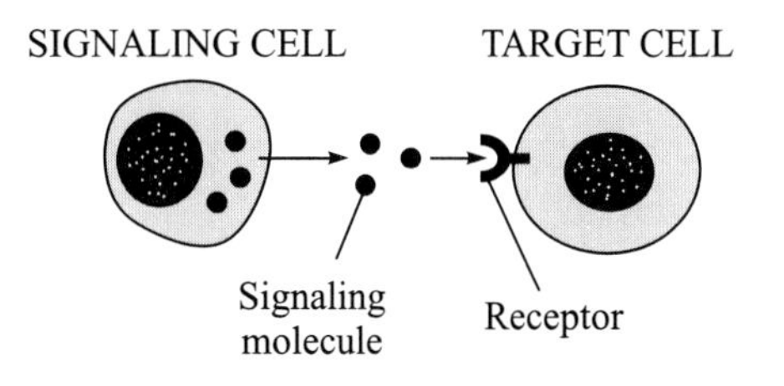

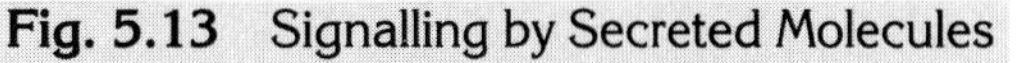

Fig. 5.13 Signalling by Secreted Molecules

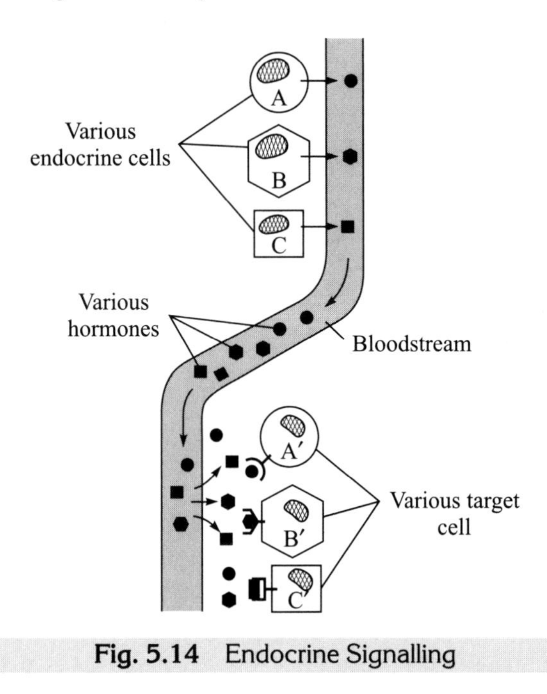

Fig. 5.14 Endocrine Signalling

Synaptic

Example - Nerve cells (or) neurons. The neuron sends electrical impulses (action potential) along its axon. When the impulse reaches the nerve terminal at the end of axon, it stimulates the terminals to secrete a chemical signal called a neurotransmitter. These neurotransmitters rapidly and specifically act and this is called as synaptic signalling.

The **contrast** between endocrine and synaptic signalling is that synaptic signals cover long distances by electrical impulses that travel at rates of upto 100 meters per second. Endocrine signalling is through transportation by blood and is comparatively slower than the electrical impulses exhibited by the synaptic signalling. The speed and specificity are unique of synaptic signalling in nerve cells or neurons.

4. Autocrine Signalling can Co-ordinate Decisions by Groups of Identical Cells

In autocrine signalling, the cell secretes a signalling molecule and it binds back to its own receptor. For effective communication, it has been found out that group of identical cells secreting signalling molecules at the same time and the receptors receiving them shows strong signalling. This is also called 'community-effect'. Example – Eicosanoids

Eicosanoids

Eicosanoids are signalling molecules often act in autocrine mode in mature mammals.

5. Gap Junctions Allow Signalling Information to be Shared by Neighboring Cells

These gap junctions are formed between closely apposed plasma membranes directly connecting the cytoplasms of the joined cells via narrow water filled channels. These channels allow the exchange of small intracellular signalling molecules such as Ca2+ and cAMP but not macromolecules such as protein, nucleic acid etc. These cells connected by gap junctions can communicate directly.

6. Each Cell is Programmed to Respond to Specific Combination of Signalling Molecules

Higher animals have specific combination of signalling molecules through specialized programmes for survival. When deprived, a cell will activate the suicide programs and kill itself. This is called 'programmed cell death' or 'Apoptosis'.

7. Different Cells can Respond Differently to the Same Chemical Signal

The same signalling molecules can induce different responses in different target cells. Example – Acetylcholine.

(a) Increases the rate of contraction in skeletal muscles
(b) Relaxation of heart muscles

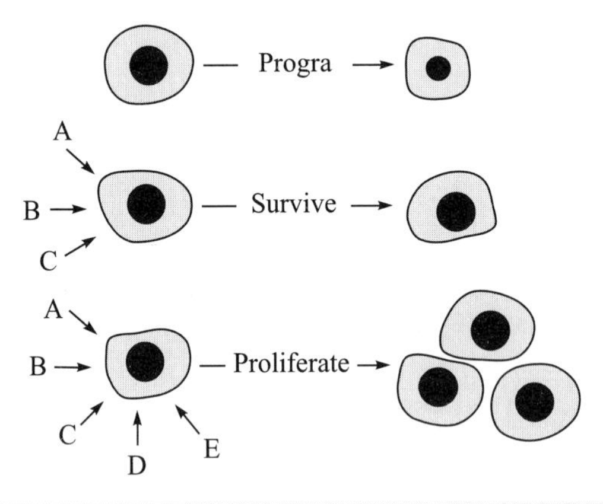

Fig. 5.15 Receptors – combinatorial Signalling

8. The Concentration of a Molecule can be Adjusted Quickly Only if the Lifetime of the Molecule is Short

When the signal is delivered and the signal is withdrawn, changes takes place in the cell

9. Nitric Acid (NO) Gas Signals by Binding Directly to an Enzyme Inside the Target Cells

Example – Nitric Oxide gas and Carbon monoxide

It has short half-life and is made up of an enzyme NO synthase. The uniqueness is that it binds to an enzyme inside the target cell. These gases not only diffuse across the membrane but these gases bind the enzyme and others to the protein.

10. There are Three Known Classes of Cell-surface Receptor Proteins

Ion channel linked, G-Protein linked and Enzyme linked classes of cell-surface receptor proteins.

Questions

1. What is a Ligand and a Receptor?
2. What are cell surface receptors?
3. Explain in detail about intra-cellular receptors:-
4. What is an effector? What is its important function? Give two examples also:-
5. What is a first and second messenger?
6. Define Signal Transduction:-
7. Define (a) autocrine, (b) paracrine and (c) endocrine system and functions:-
8. What is a signalling cell and a target cell?
9. What are receptors? Write in length about membrane bound receptors:-
10. What are the three classes of transduction? Explain
11. Explain the characteristics of receptors:-
12. What are Eicosanoids?
13. What is a gap junction?
14. What are G-Protein linked receptors?
15. What are Channel linked receptors?

CHAPTER 6

Signal Transduction

6.1 SIGNAL AMPLIFICATION

Signal transduction is the transfer of signals across the plasma membrane. The process of signal transduction is that the ligand binds to the receptor and the receptor is located on (or) in the cell. When the ligand becomes attached to the receptor, either it becomes activated or inactivated. During both these processes, the effectors recognize the message encoded in the 'ligand receptor complex' and relays to various sites within the cells by means of generating the second messenger and in addition, greatly, amplifies the signal. These collective networks is called the signal transduction.

Signal Amplification

The second messenger, in addition to relaying the information to intracellular region, strongly amplifies the signal by adding more strength to it. Signal amplification is defined as the amplification of signals received at the cell surface such as arrival of a protein hormone, growth factor, etc at the receptors through the generation of 'second messengers'.

Second Messengers

Second messenger mimics the action of hormones (extra-cellular signalling molecules) inside a cell and so they are called as second messenger.

Examples of second messengers that serve for signal amplification process

I. Cyclic AMP (cAMP)
II. Cyclic GMP (cGMP)
III. Inositol Triphosphate (IP_3)
IV. Diacylglycerol (DAG)
V. Protein kinases
VI. Calcium ions

General components of a biological system

Stimulus : Any detectable change in the system. Example – change in temperature, K+ concentration, pressure, etc.

Receptor : Protein that identifies the changes caused by the stimulus
Afferent pathway : The pathway traveled by the signal between the receptor and the integrating center
Integrating center : Signal relaying center
Efferent pathway : Pathway between integrating center to the effectors

Response

The end result triggers changes (or) sequence of changes in the target cells. The enzyme acts on another enzyme causing geometric increase causing amplification.

Cyclic AMP in signal amplification (cAMP) cAMP is 3', 5' – adenosine monophosphate, simply called cAMP, is the byproduct of ATP while activated by an enzymne called adenylate cyclase, located on the inner surface of the membrane.

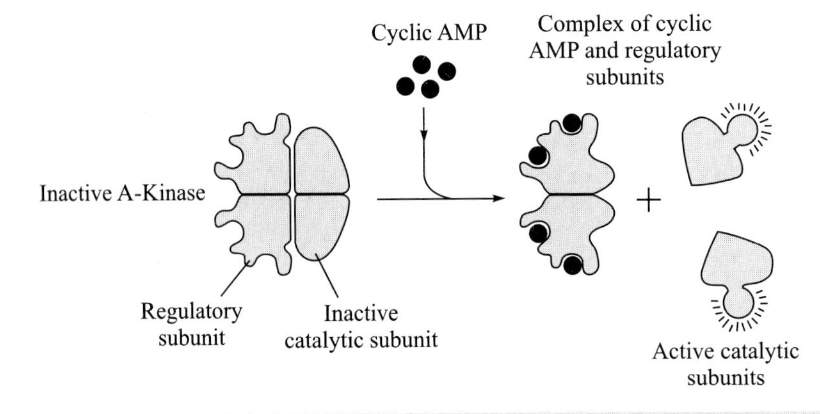

Fig. 6.1 Activation of Cyclic AMP Dependent Protein Kinase (A-Kinase)

Post receptor effects The binding of the messenger with the receptor is the only initial step leading to the cells response and creates ultimate change by any of these ways such as.

a) Membrane permeability, transport (or) electrical state,
b) The rate at which the particular substance is synthesized (or) secreted and
c) The rate or strength of muscle contraction

All changes that occur after messenger-receptor binding are due to alterations of particular cell proteins.

Some examples of Messenger-induced Responses are

a) Changes in muscle contraction – contractile proteins,
b) Changes in the rate of secretion of glucose by the liver enzyme and
c) Generation of electric signals in nerve cells – membrane proteins.

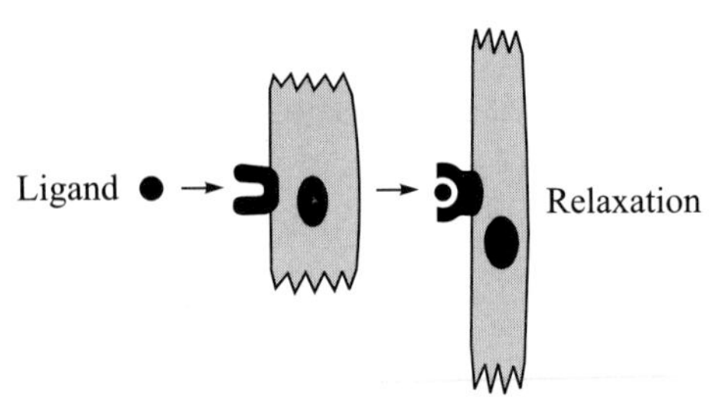

Fig. 6.2 Contraction and Relaxation in a Heart Muscle Cell

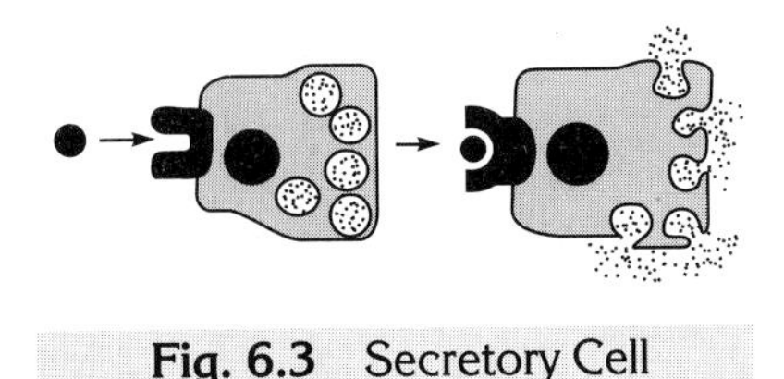

Fig. 6.3 Secretory Cell

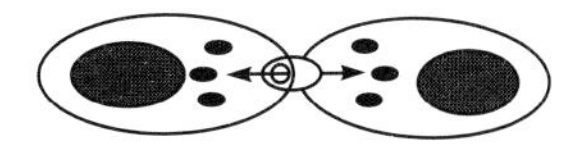

Fig. 6.4 Signalling via Gap Junctions (after Alberts *et al.*, 1989)

6.2 SIGNALLING VIA G-PROTEIN LINKED CELL-SURFACE RECEPTORS (CYCLIC AMP OR cAMP)

Adjacent to the plasma-membrane receptors are the membrane proteins which act as ion-channels through the membrane. The alteration in the membrane-receptor structure produced by the binding of the messenger permits the receptor to interact with the adjacent ion-channel protein.

a) The effect of the interaction with receptor is to open (or) close the channel, which results in an increase (or) decrease in the diffusion of the ions, specific to that channel, across the plasma membrane. Such channels are known as receptor-operated channels.
b) The ions that flow through these channels generate electric signals in nerve, muscle and some gland cells (altered membrane electric signal)

The binding of messengers to certain plasma membrane receptors causes the receptors to activate the enzyme called adenylate cyclase, which is located on the inner surface of the membrane

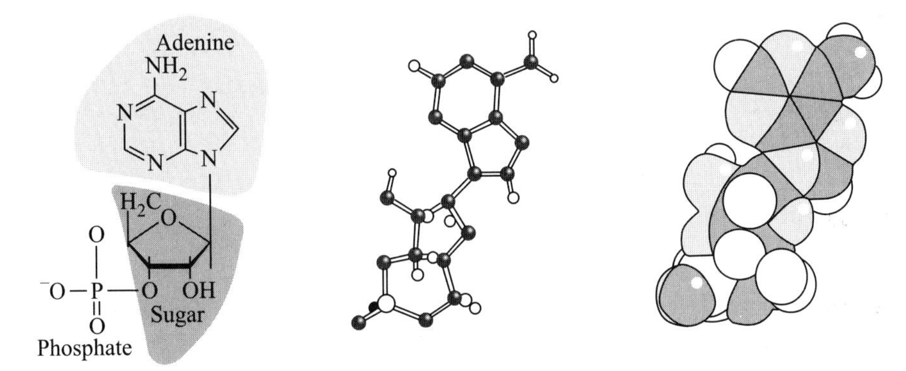

Fig. 6.5 Cyclic AMP (shown as a formula, a ball and stick model, and a space filling model – C,H,N,O and P indicate carbon, hydrogen, nitrogen, oxygen and phosphorus atoms, respectively.)

The activated adenyl cyclase catalyses intracellular ATP into 3', 5' – adenylate mionophosphate called cAMP

cAMP, then diffuses throughout the cell to make intracellular sequence of changes in the cell, leading ultimately to the cell's response. The action of cAMP is terminated by its breakdown to non-cyclic AMP, a reaction catalysed by an enzyme called phosphodiesterase. This cAMP in its intracellular space, activates one (or) more of the enzymes, collectively called as cAMP -dependent protein kinases which may be free in the cytoplasm (or) plasma membrane.

Inactive protein kinases ***gets converted into*** Active protein kinases

It is not that through the ion-channels, only ions flow through but calcium also diffuses and increases the cytosolic calcium level.

With these diffused calcium, a response is bought in a cell. This is by the binding of this diffused calcium to the particular protein found in all the cells known as calmodulin. These binding bring ultimate response in a cell (contraction, secretion etc.)

Cytosolic calcium itself is a second messenger. There are many calcium channels in the plasma membrane, which are not directly operated by the receptors. They open (or) close in response to membrane electric signals.

cAMP has influence over cytosolic calcium to alter certain proteins leading to response in a cell.

Signal Amplification by cAMP

Activation of adenylate cyclase (enzyme) (or) protein kinase initiates a 'cascade' of events. Let us say, for example, cAMP has 100 numbers of molecules and active protein kinase has 00 numbers of molecules and then 100 x 100 = 10000 is active protein – phosphate enzyme and the final amplification is 10000 x 10000 = 1,00,00,000.

6.3 BIOSYNTHESIS AND ROLE OF INOSITOL- TRIPHOSPHATES AS MESSENGERS

Inositol phosopholipid is a minor phospholipid present in the cell membrane. When an extracellular signalling molecule binds to the G protein linked receptor present in the cell surface, this inositol phospholipid is activated and changes into a different form, Phosphatidylinositol (PI). The Inositol Phospholipids are very important in signal transduction and particularly, the phosphorylated dervivatives of PI namely PI phosphate (PIP) and PI biPhosphate (PIP_2), that are located in the inner half of the plasma membrane lipid bilayer. This $PIP_2$is less available in cell membranes than PI, but the hydrolysis of PIP_2 is very important for signal transduction.

Phosphorylation of Inositol Phospholipid

The chain of events leading to PIP_2 breakdown begins with the binding of a signalling molecule to a G protein linked receptor in the plasma membrane. An activated receptor stimulates a trimeric G protein called 'Gq'. Gq activates a Phosphoionosite-specific phospholipase C called Phspholipase C - β. In less than a second, this enzyme cleaves PIP_2 and generates two products namely, Inositol triphosphate and DiAcyl Glycerol.

Some cellular responses mediated by G-protein linked receptors coupled to the Inositol-Phospholipid signalling pathway

Target Tissue	*Signalling Molecule*	*Major Response*
1. Liver	Vasopressin	Glycogen breakdown
2. Pancreas	Ach	Amylase secretion
3. Smooth muscle	Ach	Contraction
4. Mast cells	Antigen	Histamine secretion
5. Blood platelets	Thrombin	Aggregation

Inositol Phospholipids & Hydrolysis of PIP_2

Two intracellular mediators are produced as a result of hydrolysis of PIP_2. They are Inositol Triphosphate (IP_3) and DiAcylGlycerol (DAG). IP_3 diffuses through the cytosol and releases Ca^{2+} from the ER. DAG remains in the membrane and helps to activate the enzyme called Protein Kinase C. There are three classes of phospholipase. They are C-β, γ and δ. C-β class of phospholipase is activated by G-protein linked receptors. γ class is activated by a second class of receptors called tyrosine kinase. This tyrosine kinase activates IP signalling pathway without any intermediary G protein.

Inositol Triphosphate (IP_3) – Release of Ca^{2+} from ER

The Inositol TriPhosphate (IP_3) is produced by PIP_2 (PhosphatidylInositol Biphosphate) hydrolysis and is a small water soluble molecule that leaves the plasma membrane and diffuses rapidly through the cytosol. In the cytosol, it releases Ca^{2+} from ER by binding to IP_3 – gated Ca^{2+} release channels in the ER membrane. These Ca^{2+} channels are similar to ion channels. Two mechanisms operate to terminate the initial Ca^{2+} response.

A. IP_3 is rapidly dephosphorylated (inactivated) by specific phosphatases

B. Ca^{2+}, that enters the cytosol is rapidly pumped out of the cell.

All of the IP_3 are not dephosphorylated. However, some phosphorylate to form Inositol 1,3,4,5 – tetrakisphosphate (IP_4). IP_4 mediate slower and more prolonged responses in a cell.

Ca^{2+} oscillations often Prolong the initial IP_3 – induced Ca^{2+} Response

When Ca^{2+} sensitive fluorescent indicators such as aequorin (or) Fura-2 are used to monitor cytosolic Ca^{2+}in individual cells in which the Inositol phospholipid signalling pathways has been activated, the Ca^{2+}signal is often seen as a wave. The transient increase in Ca^{2+} is often followed by a series of Ca^{2+} spikes each lasting for seconds (or) minutes and these spikes are called as Ca^{2+} oscillations. These Ca^{2+} oscillations can persist for as long as the receptors are activated in the plasma membrane. The biological significance of this Ca^{2+} oscillation is uncertain and their frequency depends on the concentration of the Extra cellular signalling molecules or ligand.

DiAcylGlycerol (DAG) Activates Protein Kinase (C-Kinase)

DiAcylGlycerol (DAG) is a product of hydrolysis of PIP_2. The other product IP_3 increases the concentration of Ca^{2+}in the cytosol. DiAcylGlycerol has two potential signalling roles, namely,

A. It can further be reduced to release Arachidonic Acid. This Arachidonic acid can either act as a messenger (or) can be used in the synthesis of Eicosanoids.

B. DiAcylGlycerol activates serine / Threonine Protein Kinase that phosphorylates slected proteins in the target cell.

The enzyme activated by DiAcylGlycerol is called Protein Kinase C (C-Kinase) or PKC because it is Ca^{2+} dependent. DiAcylGlycerol cannot sustain the activity of C-Kinase for long term responses such as cell proliferation (or) differentiation. For sustaining prolonged responses, DiAcylGlycerol secretes phosphatidylcholine. When activated C-Kinase phosphorylates specific Serine / Threonine recidues on target proteins depending on the cell type. Highest concentration of C-Kinase is found in brain and in many cells, the activation of C-Kinase increases the transcription of specific genes takes place.

Examples

Two known pathways

(a) Inactive C-Kinase (No transcription of genes 1 and 2)
(b) Active C-Kinase (Activated Transcription of genes 1 and 2

(a) In inactive C-Kinases, no transcription of genes 1 and 2 takes place.

- Plasma membrane
- Cytoplasm or cytosol

⇓

Inactive C-Kinase, MAP kinase, Ik-B (Inhibitor protein) and NF-kB (regulatory protein)

- Nucleus

⇓

SRE (Serum Response Element)
SRF (Serum Response Factor)
EIK-1 (Protein)

(b) Active C-Kinase

One pathway

C-Kinase
⇓ *Phosphorylation*
MAP kinase
⇓ *Phosphorylates*
EIK-1
⇓
EIK1 + SRF
⇓ *Activates*
SRE
⇓
mRNA ⇒ **Gene 1**

(c) Another Pathway

C-Kinase
⇓ *Phosphorylation*
Ik-B
⇓ *Releases*
NF-kB
⇓ *Active transcription (nucleus)*
mRNA ⇒ **Gene2**

In both pathways, Ik-B (inhibitor protein) or regulatory protein stimulates the transcription of specific genes.

Conclusion

All extracellular signals are greatly amplified by the use of intracellular mediators and enzymatic cascades. Amplification is the same for Inositol-Phospholipid molecules like cAMP.

6.4 SIGNALLING VIA ENZYME-LINKED LINKED CELL-SURFACE RECEPTORS (CYCLIC GMP OR cGMP) CYCLIC GMP (OR) cGMP

Cyclic GMP, cGMP is an enzyme linked receptor (3', 5' – Guanaosine monophosphate), activated by an enzymne called Guanylate cyclase, that catalyses the production of cGMP in the cytosol.

Role of cGMP

a) cGMP is synthesized from ATP using an enzyme, guanylate cyclase. cGMP serves as a second messenger for Nitric Oxide (NO)
b) ANP (Atrial Natriuretic Peptide)
c) The response of the rods of the retina to light
d) Smooth muscle tone
e) Water balance and ion fluxes
f) Neuronal plasticity

Guanyl Cyclase

The activation enzyme, guanyl cyclase has two families

1. Cytoplasmically localized soluble Guanyl cyclases (or) sGC's
2. Membrane associated receptor Guanyl cyclases (or) rGC's

Cytoplasmically localized soluble guanyl cyclases (or) sGC's

It is a hetero dimeric protein and helps in the binding of a heme prosthetic group and can be activated by free radical messengers such as Nitric Oxide, NO (Inter and Intra cellular)

Membrane associated receptor guanyl cyclases (or) rGC's

This is activated by extracellular ligands, usually peptide hormones (binding of the extra-cellular protion of the protein)

Generation of cGMP

A. Adjacent to the plasma-membrane receptors are membrane proteins that act as ion-channels through the membrane. The alteration in the membrane – receptor atructure produced by the binding of the messenger permit the receptor to interact with the adjacent ion-channel protein.
B. The binding of the messenger to certain plasma-membrane receptors causes the receptors to activate the enzyme called guanyl cyclase which consists of two major sub-families
   a) sGC's (soluble Guanyl cyclases)
   b) rGC's (receptor bound Gunayl cyclases)

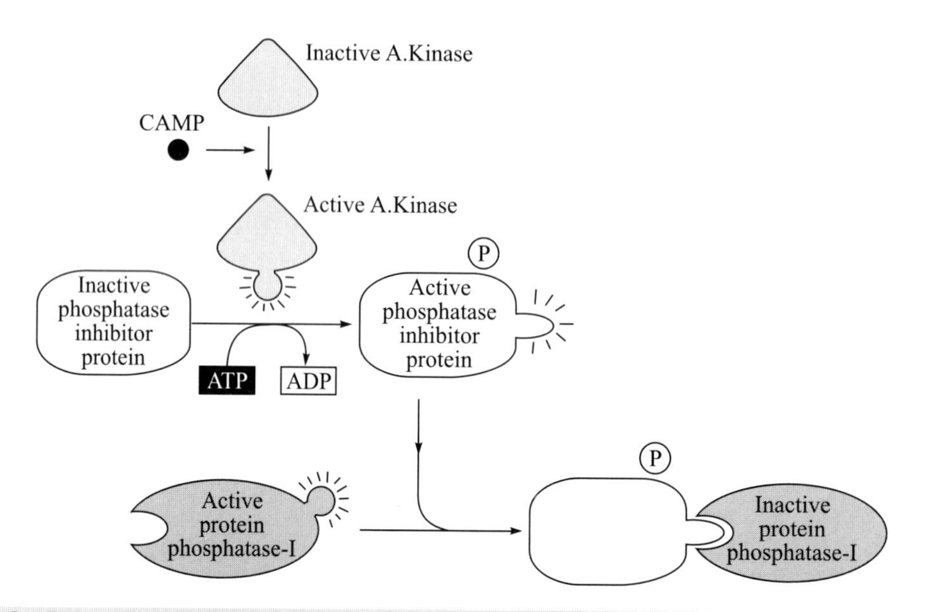

Fig. 6.6 Role of Protein Phosphatase –I in the Regulation of Glycogen Metabolism by Cyclic AMP

C. sGC's becomes activated by the NO in the cytoplasm and catlyses intracellular ATP into 3', 5' – Guanaosine monophosphate called as cGMP

D. rGC's becomes activated when the messenger binds to the receptor and catalyses intracellular ATP into 3', 5' – Guanaosine monophosphate as cGMP.

E. cGMP, produced by sGC's diffuse throughout the cell to make intracellular sequences of changes in a cell, leading ultimately to the cell's response

F. The action of cGMp activates inactive protein kinase into active protein kinase G. Protein Kinase ***gets converted by sGC's into*** Protein Kinase G

G. The action of cGMP is terminated by its breakdown to non-cyclic GMP, a reaction catalysed by an enzyme called phosphodiesterase

H. Receptor-Protein – Rhodopsin of the human eye is a very good example to be illustrated.

Example

The Working of Retina

How the retina works in a human eye

Light rays from an object → Cornea → Lens (where refraction takes place) → Retina (inverted image)

At the retina → light is converted into electrical signals, and

Electrical signals are carried by optic nerve → Brain

Light comes to the retina through the retinal blood supply and the light also passes through two transport layers of neurons namely the ganglion cells and the bipolar cells. Then, it finally reaches the cones and rods, which convert light into an electric impulse. There are 100 million rods and cones on the retina of each eye and these rods are sensitive to dim or bright and insensitive to colour whereas the cones are sensitive to colour and insensitive to dim or bright light.

Rod Cells

In the rod cells, the receptor protein is Rhodopsin and this is a seven pass transmembrane protein.

Activated Rhodopsin stimulates

↓

Transducin (G protein)

↓

Activates

↓

rGC's enzyme

↓

cGMP

↓

cGMP activates rods and keeps the dark channel closed

↓

Movement of Na^+ and Ca^+ ions

↓

sGC enzyme

↓

Na^+ and Ca^+ ion movement stops →

↓

Bright

Signal Amplification by cGMP

Activation of the rGC's (or) sGC's triggers a cascade of cellular reactions resulting in the amplification of a single signal. The example of the activation of rhodopsin explains the activation pattern and amplification of a signal.

+ (one rhodopsin molecule absorbs one photon)

↓

500 transducin molecules are activated

↓

10^5 cyclic GMP molecules are hydrolysed

↓

250 Na^+. Ca^+ channels close / open

↓

Result : Amplification by rods / cones of the retina of the human eye

6.5 ROLE OF G PROTEINS IN SIGNAL TRANSDUCTION

G Proteins

G proteins are one of the large families of heterodimeric GTP binding protein that serve as an important intermediate in cell-signalling pathways.

Signal Transduction Process

G protein linked receptors initiate a cascade of chemical events within the target cell that alters the concentration of small intracellular messengers such as cAMP (or) inositol triphosphate. These intracellular messengers in turn alter the behavior of other intracellular proteins. **cAMP** levels affect cells by stimulating cAMP dependent **protein kinase** to phosphorylate specific target proteins. **Calcium levels** modify the activity of certain enzymes by binding to the calcium binding protein called **calmodulin**. The effects of all these second messengers are rapidly reversible when the extracellular signal is removed. The response of cells to external signals initiates signalling cascades that can greatly amplify and regulate various inputs.

Uniqueness

G protein has a seven pass tyransmembrane (7 – trans-pass) spanning serpentine receptors and perform a lot of functions such as

a) Olfactory bulbs in the nose
b) Odorant ligands
c) Rods and cones in the retina
d) Yeast cells (releases a polypeptide mating factor and it serves as a ligand and this follows autocrine mode of signalling)

Examples

A. Epinephrine

This example involves the ligand, epinephrine that is also known as adrenaline that is released by the adrenal glands. The adrenal glands are situated above the kidneys in response to very stressful stimuli. Once released, this epinephrine passes throughout the blood stream and adsorbs to specific receptors on the surface of cells in various tissues throughout the body.

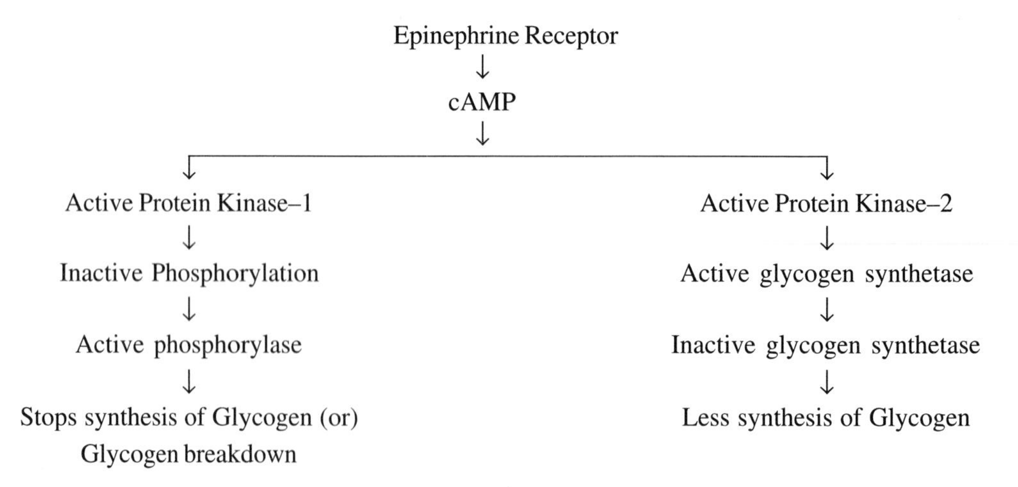

The β-adrenergic receptor is also a 7-membarne spanning, serpentine receptor ambedded in the plasma membrane of the cells. This membrane bound β-adrenergic receptor is bound to hold the signal only for a short period. This receptor receives the signal from the specific active G protein and in turn is released into the cytoplasm. The β-adrenergic receptor communicates with the cytoplasm by stimulating a second protein called G1 (or) G protein itself. This second protein, herein, also called as G protein remains in an inactive state in the cytoplasm.

Cytoplasmic Signal Transduction

The second protein, G, remains in an inactive state. When ligand binding by another G protein activates this protein, it switches itself into an active state. Once it becomes active, the G protein will send signals further into the cell. However, the activated G protein will remain active only for a short period of time, after which it will shut itself off. G proteins act as a binary switch – proteins will exist in two states, 'ON' or 'OFF'. This function of the switch is decided by the nucleotide with which it binds. When it is inactive, it binds to GDP (Guanosine Di Phosphate) and when it is active, it binds to GTP (Gunaosine Tri Phosphate). GDP always is found in a bound state. Bound state means Trimeric proteins namely α,β and γ staying intact in a bound manner. GTP disfigures the configuration and α gets separated fromβ and γ. Therefore, β and γ always stays together while GTP is active and α alone.

Example

Rhodopsin

Rhodopsin is the first member of the family of 7-helix receptors to have its structure determined by X-ray crystallography. This 7-helix receptor interacts with a G-protein, a hetero trimeric GTP binding protein. The three subunits of a G protein are designated as α,β and γ. The seven helix receptors that interact with the G proteins are called GPCR (or) G protein Coupled Receptors. Various proteins interact with GPCRs to modulate their activity. G protein, that is part of an pathway that stimulates adenylate cyclase is called Gs and its α subunit as G sα. G proteins α subunit (Gα) binds to GTP and can hydrolyze it to GDP and Pi. The complex of β and γ subunits inhibits Gα.

Sequence of Events in Amplification

On – signal

a) Initially, Gα has bound with GDP and therefore, α,β and γ subunits are complexed together.
b) When a 7-helix receptor (GPCR) binds with a G protein, Gα leaves GDP and binds to GTP. This is called as GDP-GTP Exchange.
c) GDP-GTP exchange causes another conformational change in Gα
d) GDP-GTP dissociates from the β - γ complex and binds to activate adenylate cyclase.
e) Adenylate cyclase, activated by Gα-GTP catalyses synthesize cAMP.
f) As a result, protein kinase – A (cAMP dependent protein kinase) catalyses phosphorylation of various cellular proteins.

Off – signal

g) Gα hydrolyzes GTP to GDP and Pi
h) Gα rebinds β - γ complex
i) Adenylate cyclase is no longer activated
j) Phosphodiesterase catalyses hydrolysis of cAMP into AMP

Signal Amplification

One hormone molecule can lead to the formation of many cAMP molecules and each catalytic subunit of protein kinase A phosphorylates many proteins.

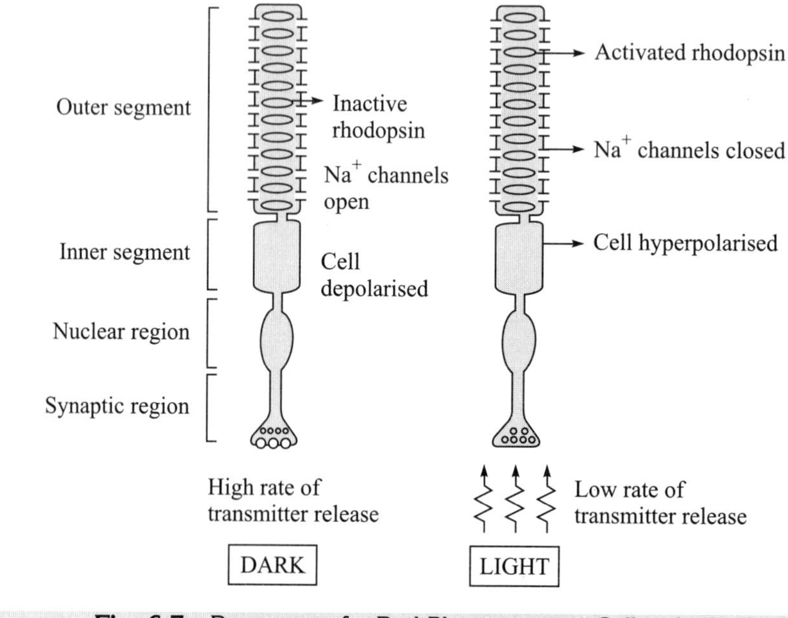

Fig. 6.7 Response of a Red Photoreceptor Cell to Light

Example

The example of the activation of rhodopsin explains the activation pattern and amplification of a signal.

+ (one rhodopsin molecule absorbs one photon)

↓

500 transducin molecules are activated

↓

10^5 cyclic GMP molecules are hydrolysed

↓

250 Na^+. Ca^+ channels close / open

↓

Result : Amplification by rods / cones of the retina of the human eye

↓

The heterodimeric G proteins activates a second protein called Transducin

Some Other GTP Binding Proteins

- Growth factors

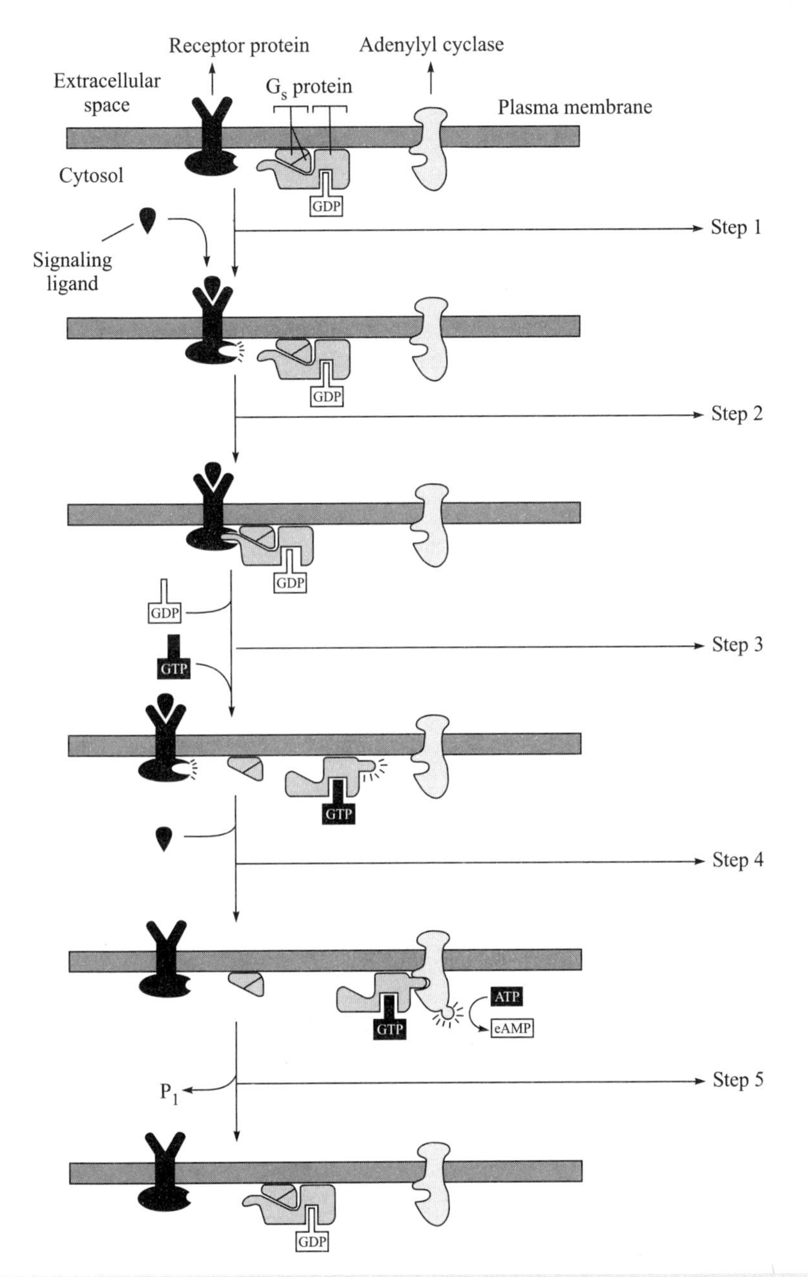

Fig. 6.8 A Current Model of How Gs Couples Receptor Activation to Adenylyl Cyclase Activation

Step – 1 : Ligand binding alters conformation of receptor, exposing binding site for Gs Protein.
Step – 2 : Diffusion in the bilayer leads to association of ligand-receptor complex with Gs protein, thereby greatly weakening the affinity of Gs for GDP
Step – 3 : GDP dissociates, allowing GTP to bind; this causes the a subunit to dissociate from the Gs complex, exposing its binding site for adenylyl cyclase
Step – 4 : The a subunit binds to and activates adenylyl cyclase to produce many molecules of camp; meanwhile, dissociation of the ligand returns the receptor to its original conformation
Step – 5 : Hydrolysis of the GTP by the a subunit returns the subunit to its original conformation, causing it to dissociate from the adenylyl cyclase (which becomes active) and to reassociate with bg complex to re-form Gs

- Ras
- Rab
- ARF
- Ran
- Rho

6.6 CALCIUM ION FLUX AND ITS ROLE IN CELL SIGNALLING

Usually, cytosolic calcium is kept low by Ca++-ATPase pumps in plasma membrane and endoplasmic reticulum (ER) membranes. These members of the P-class (Ca++) family of ion pumps transport Ca++ away from the cytosol, either out of the cell or into the ER.

Ca++ Concentrations

Cytosolic Ca++ is usually in the μM (10^{-6}) range. The extracellular Ca++ in mammals is in the mM (10^{-3}) range. Ca++ is also relatively high within the ER lumen which serves as a reservoir for Ca++. Mitochondria also serve as a reservoir for Ca++.

Ca++ Binding Proteins

Ca++ binding proteins in the ER lumen "buffer" free Ca++ and increase the capacity for Ca++ storage. ER Ca++ binding proteins have 20-50 low affinity Ca++ binding sites per molecule, consisting of acidic residues.

Calsequestrin is in the lumen of sarcoplasmic reticulum (SR), the specialized ER of muscel

Calreticulin is in the lumen of the ER of non-muscle cells. It also has a chaperone role.

Calmodulin is found in the cytosol of all cells in inactive form

Calcium Imaging

Cytosolic Ca++ may be monitored using indicator dyes or proteins that are either luminescent or change their fluorescence when they bind Ca++. Fluorescent indicators used with confocal fluorescence microscopy can provide high-resolution imaging and quantitation of Ca++ fluctuations within cells.

Signal Activated Ca++ Channels

A signal cascade may activate Ca++ channels in ER or plasma membranes allowing Ca++ to rapidly enter the cytosol. Any increase in cytosolic Ca++ is usually transient because Ca++-ATPases pump Ca++ out of the cytosol. A transient increase in cytosolic Ca++ may be localized to the vicinity of a Ca++ release channel. Such a localized Ca++ puff may activate effectotrs that induce additional Ca++ release, giving rise to a more widely distributed increase in cellular Ca++ that may spread to neighboring cells. Pulsatile Ca++ waves may also occur.

Ryanodine Receptor

A large Ca++ release channel in the membrane of muscle sarcoplasmic reticulum (SR) is called the ryanodine receptor, because of sensitivity to inhibition by a plant alkaloid ryanodine. Skeletal muscle contraction is activated when Ca++ is released from the SR lumen to the cytosol via the ryanodine receptor.

Tubules

Invaginations of muscle plasma membrane are called as T tubules. Voltage gated Ca++ channels in the T tubule membrane interact with ryanodine receptors in the closely apposed SR membrane. Activation of voltage-gated Ca++ channels, by an action potential in the T tubule, leads to opening of ryanodine – sensitive Ca++ release channels. Ca++ moves from the SR lumen to the cytosol, passing through the transmembrane part of the ryanodine receptor and then through recpetor's cytoplasmic assembly. The ryanodine receptor is itself activated by cytosolic Ca++, leading to amplification of the Ca++ signal.

Inositol Tri Phosphate triggers Ca++ release

In some mammalian cells, IP_3, Inositol TriPhosphate, triggers Ca++ release from the ER. The second messenger, IP_3 is produced – example – in response to hormonal signals, from the membrane lipid phosphatidyl inositol. IP_3 activated Ca++ release channels are examples for ligand-gated channels and helps the Ca++ binds to a protein called calmodulin.

Many cellular reactions & processes are regulated by Ca++

Many cellular reactions and processes are regulated by Ca++. Calmodulin, a Ca++ activated switch protein, mediates many of the signal functions of Ca++. Calmodulin co-operatively binds 4 Ca++ and at each binding site, Ca++ interacts with oxygen atoms, of Glutamine and Aspartite side chain carboxyl groups and of a protein backbone, in a loop between 2 alpha helices at right angles. This helix-loop-helix motif is called EF hand.

Helix – Loop – Helix in Calmodulin

There are 4 helix-loop-helix motifs in calmodulin. In the absence of Ca++, calmodulin assumes a dumbbell shape, with 2 of the helix-loop-helix motifs at each end of the molecule. Ca++ binding allows calmodulin to interact with various target proteins. Activated calmodulin wraps around the target domain of a calmodulin-sensitive protein, altering that protein's activity. The target domain is a positive charged, amphipathic α-helix, with polar and non-polar surfaces. Met residues of calmodulin bid hrdrophobic residues in target domains of enzymes regulated by calmodulin.Some protein kinases that transfer phosphate from ATP to hydroxyl residues on enzymes to be regulated, are activated by Ca++ calmodulin.

Signal turn-off

The plasma membrane Ca++-ATPase that pumps Ca++ out of the cell, is one of the target proteins activated by Ca++-Calmodulin. This itself turns off Ca++ signals.

6.7 PHOSPHORYLATION OF PROTEIN KINASE

cAMP exerts its effect in animal cells mainly by activating the enzyme called cAMP dependent protein kinases (or) A-Kinase. cAMP mediated protein phosphorylation was first demonstrated in studies of glycogen metabolism in skeletal muscle cells. Glycogen breakdown is demonstrated with reference to skeletal muscle and protein kinase – A (PKA).

Example –1

Epinephrine receptor

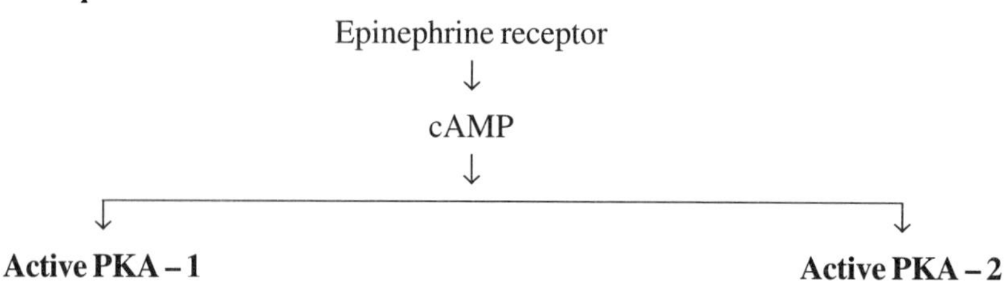

Active PKA – 1: Inactive Phosphorylases into active phosphorylase thereby catalyses glycogen breakdown.

Active PKA – 2: Active Glycogen Synthetase gets converted into Inactive glycogen synthetase thereby synthesizing less amount of glycogen.

Inactive Kinase A with the activation of cAMP activates A-Kinase and as a result inactive phosphorylase kinase becomes active phosphorylase. This activated phosphorylase kinase hydrolysis forming ATP into ADP and glycogen is broken into glucose-1-phosphate, the process being called as glycolysis, leaving inactive glycogen

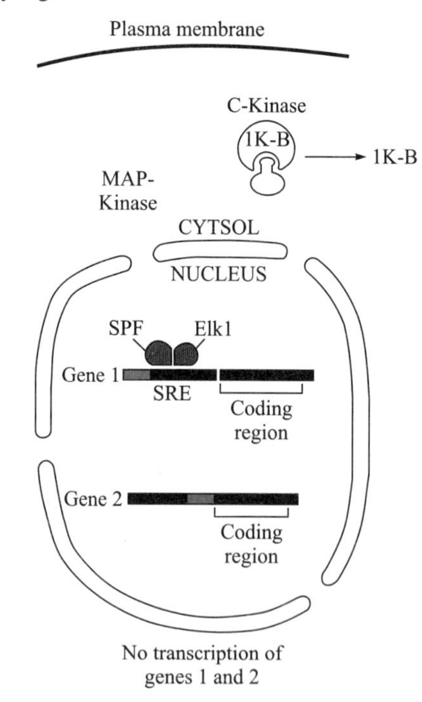

Fig. 6.9 Two Intracellular Pathways by which Activated C-Kinase can Acivate the Transcription of Specific Genes

Example – 2

Role of protein Phosphatase – I in the regulation of Glycogen metabolism by cAMP

Inactive A Kinase ⇒ cAMP ⇒ Active A Kinase ⇒ gets converted into (1) Inactive phosphatase inhibitor protein and (2) active phosphatase inhibitor protein thereby releasing ATP and ADP. The activation phase in turn makes the active protein phosphatase – I into inactive protein phosphatase – I.

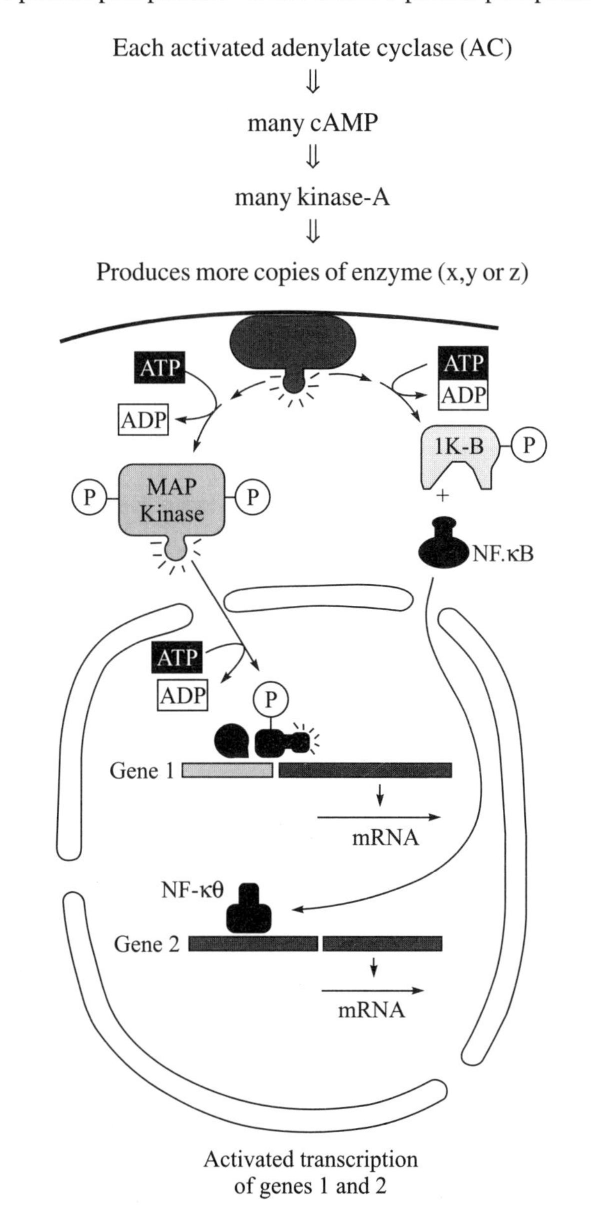

Fig. 6.10 Two Intracellular Pathways by which C-Kinase can Activate the Transcription of Specific Genes

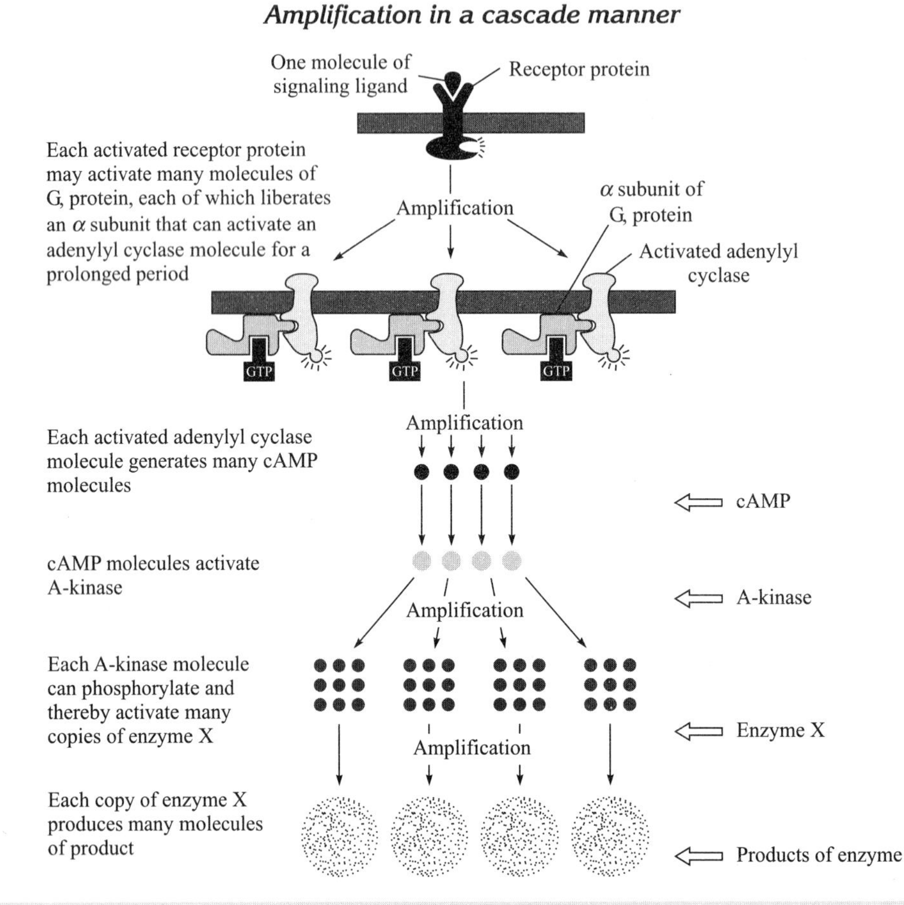

Fig. 6.11 Amplification Process taking Place in a Cascade Manner (after Alberts *et al.*, 1989)

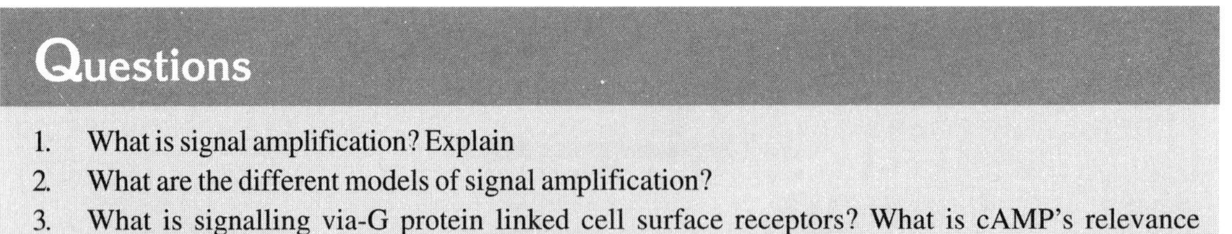

Questions

1. What is signal amplification? Explain
2. What are the different models of signal amplification?
3. What is signalling via-G protein linked cell surface receptors? What is cAMP's relevance regarding this?
4. Explain in detail about the biosynthesis and role of Inositol triphosphate as messengers:-
5. Write an essay on the signalling via enzyme linked cell surface receptors (cGMP):-
6. What are the significant issues of G Protein in signal transduction?
7. What is the role of Calcium flux in cell signalling?
8. What is Calmodulin? What is its role in the activation of DiAcylGlycerol?
9. What are the current models of signalling and amplification?
10. What do you understand by the term 'Phosphorylation'? Explain with regard to Protein Kinase:-

CHAPTER 7

Cell Culture

7.1 INTRODUCTION

Cell culture has become a vital technology in various branches of life sciences. Cell culture basically involves the maintenance and propagation of cells, both prokaryotic and eukaryotic in a suitable nutrient media (*in vitro* in the laboratory). Thus, culturing is a process of growing cells artificially.

Cell culture evolved first when Ross Harrison during 1907 performed a frog tissue culture technique. The choice of selection of frog might be for two reasons: being a cold-blooded animal, no incubation is required and tissue regeneration is fast in frog. Subsequently, during 1940s chick embryo tissue culture techniques was developed and during 1950s, human tissues were demonstrated by He La. He La observed that human tumor cells could give rise to continuous cell lines. Among the various animal cell cultures, mouse cell cultures are the most commonly used in the laboratory.

Prokaryotes are organisms without a cell nucleus, or indeed any other membrane-bound organelles, in most cases unicellular (in rare cases, multicellular). This set of characteristics is distinct from eukaryotes, organisms that have cell nuclei and may be variously unicellular or multicellular. The difference between the structure of prokaryotes and eukaryotes is so great that it is considered to be the most important distinction among groups of organisms.

Prokaryotes also lack cytoskeletons and do lack membrane-bound cell compartments such as vacuoles, endoplasmic reticula, mitochondria and chloroplasts. Prokaryotes have a single circular chromosome, contained within a region called *nucleoid*, rather than in a membrane-bound nucleus, but may also have various small circular pieces of DNA called plasmids spread throughout the cell. Prokaryotes also have cell walls, while some eukaryotes, particularly animals, do not. Both eukaryotes and prokaryotes have structures called ribosomes, which produce protein. Prokaryotes are usually much smaller than eukaryotic cells.

A **eukaryote** is an organism with a complex cell or cells, in which the genetic material is organized into membrane-bound nuclei. Eukaryotes comprise animals, plants, and fungi–which are mostly multicellular–as well as various other groups that are collectively classified as protists (many of which are unicellular). Eukaryotic cells are generally much larger than prokaryotes, typically a thousand times by volume. They have a variety of internal membranes and structures, called organelles, and a cytoskeleton composed of microtubules and microfilaments, which play an important role in defining the cell's organization. Eukaryotic DNA is divided into several bundles called chromosomes, which are separated by a microtubular spindle during nuclear division.

Important Terminologies

The term cell culture is commonly used to include both organ culture and cell culture. Some of the **important terminologies** are defined as follows:

Organ culture: The culture of native tissue (i.e. undisaggregated tissue) that retains most of the vivo histological features is regarded as an organ culture.

Cell culture: This refers to the culture of dispersed (or disaggregated cells obtained from the original tissue or from a cell line.

Histotypic culture: The culturing of the cells for their segregation to form a tissue-like structure represents histotypic culture.

Organotypic culture: This culture technique involves the recombination of different cell types to form a more defined tissue or an organ.

Primary culture: The culture produced by the freshly isolated cells or tissues taken from an organism is the primary culture. These cells are heterogeneous and slow growing and represent the tissue of their origin with regard to their properties.

Cell line: The sub culturing of the primary culture gives rise to cell lines. The term continuous cell line implies the indefinite growth of the cells in the subsequent sub culturing. On the other hand, **finite cell lines** represent the death of cells after several subcultures.

7.2 TECHNIQUES FOR THE CULTURE OF PROKARYOTIC AND EUKARYOTIC CELLS

Techniques to culture vary widely in terms of growth media, pH and temperature, be it either prokaryotic or eukaryotic cell.

A growth medium is an object in which cells experience growth. There are different kinds of media for growing different sorts of cells. The difference in growth media are between those used for cell culture and for specific cell types derived from plants or animals, and microbial culture used for growing microorganisms, usually bacteria or yeast. These differences arise due to the fact that cells derived from whole organisms and grown in culture are often incapable of growing without the provision of certain requirements, such as hormones or growth factors which usually occur *in vivo*.

In the case of animal cells these requirements are often provided by the addition of blood serum to the medium. These media are often red or pink due to the inclusion of pH indicators. In the case of microorganisms, there are no such limitations as they are often unicellular organisms. Another major difference is that animal cells in culture are often grown on a flat surface to which they attach, and the medium is provided in a liquid form, which covers the cells. Bacteria such as *Escherichia coli* may be grown on solid media or in liquid media and liquid nutrient medium is commonly called *nutrient broth*.

The most common growth media for microorganisms are nutrient broth or Luria-Bertani medium (L-B medium). Bacteria that have grown in liquid cultures often form colloidal suspensions. When agar, a substance that sets into a gel, is added to a liquid medium it can be poured into petridishes where it will solidify, called agar plates, and provide a solid medium on which microbes may be cultured.

Another important distinction between different growth media is that of a defined and an undefined medium. A defined medium will have known quantities of all ingredients, for microorganisms they consist of providing trace elements, any vitamins required by the microbe and especially a defined carbon source and nitrogen source (for example, glucose or glycerol is often used as carbon sources; ammonium salts or nitrates as inorganic nitrogen sources). Nutrient media are not defined media as they contain ingredients such as yeast extract which vary in composition depending on the source. This definition applies to cell culture media as well, where any medium containing, for example, animal blood serum is undefined, as the composition of the serum will vary from supplier to supplier and batch to batch.

Some organisms require specialized environments, for example, since viruses are obligatory intracellular parasites, they require a growth medium composed of living cells. Often special media are required for microorganism and cell culture growth. Selective media are used to grow cells which possess a desired selectable trait. For example, if a microorganism is resistant to a certain antibiotic, then that antibiotic can be added to the medium in order to prevent other cells, which do not possess resistance, from growing.

Differential media are used to distinguish one microorganism type from another growing on the same media. While selective media are used to allow the growth of only select microorganisms, differential media allow the growth of multiple types, but result in distinguishing characteristics (such as the production of a brightly colored or phosphorescent dye). Growth media can be both selective and differential.

7.3 PLANT CELL CULTURE

Plant Cells in culture can be genetically transformed by a number of techniques. Explants can be transformed directly by *Agro bacterium tumefaciens* or by bombardment with DNA-coated particles from a particle gun. Protoplasts (plant cells with their cell walls removed) are the targets of microinjection and electroporation (DNA uptake mediated by an electric field). Plant cell culture is central to these techniques in that cell culture allows transformants to proliferate and sometimes, regenerate into genetically identical clones. While many formulations for plant tissue culture media have been developed, all contain the same basic types of ingredients such as:

- Inorganic Salts (Macro and Micronutrients)
- Vitamins
- Organic Carbon Source

In addition to these components, most formulations contain plant growth regulators (plant hormones), auxin and/or cytokinin. Plant tissue culture media are generally solidified by the addition of a gelling agent. The medium utilized in this experiment contains an agar substitute, which forms a clear, colorless, strong gel that aids in the detection of microbial contamination.

Minimal Requirement for Cell Culture Infrastructure

- Clean premise with sterile area
- Preparation facilities
- Animal house
- Microbiology laboratory
- Storage facilities (for glassware, chemicals liquids and small equipment)

Equipment

Laminar flow, incubator, refrigerator and freezer (-20°C), balance. CO_2 cylinder, centrifuge, inverted microscope, water purifier, hemocytometer, liquid nitrogen freezer, slow-cooling device (for freezing cells), pipette washes and deep washing sink.

Besides the requirements listed above, there are additional facilities that may be beneficial or useful for tissue cultures. These include air-conditioned rooms, containment room for biohazard work, phase-contrast microscope, centrifuge and video equipment.

Culture Vessels

In the tissue culture technology, the cells attach to the surface of a vessel that *serves as a substrate*, and grow. Hence there is a lot of importance attached to the nature of the materials used and the quality of the culture vessels.

Anchorage-dependent cells are used when the cells require an attachment for their growth. On the other hand, some cells undergo transformation and become *anchorage independent.*

Materials Used for Culture Vessels

Glass: Although glass was the original substrate used for culturing, its use is almost discontinued now. This is mainly because of the availability of more suitable and alternative substrates.

Disposable plastics: Synthetic plastic materials with good consistency and optical properties are now in use to provide uniform and reproducible cultures. The most commonly used plastics are polystyrene, polyvinyl chloride (PVC), polycarbonate, metinex and thermones (TPX).

Types of Culture Vessels

Multi-well plates, petridishes, flasks and stirrer bottles are the common types of culture vessels. The actual choice of selecting a culture vessel depends on several factors, such as the way cells grow in culture-monolayer or suspension, quantity of the cells required, frequency of sampling for the desired work, purpose for which the cells are grown and the cost factor.

In general, for *monolayer cultures* the cell yield is almost proportional to the surface area of the culture vessel. The flasks are usually employed for the purpose. Any type of culture vessel can be used to grow suspension cultures. It is necessary to slowly and continuously agitate the suspended cells in the vessel.

Treatment of Culture Vessel Surfaces

For improving the attachment of cells to the surfaces and for efficient growth, some devices have been developed. It is a common observation that the growth of the culture cells is better on the surfaces for second seeding. This is attributed to matrix coating of the surfaces due to the accumulation of certain compounds like collagens and fibronectin released by the cells of the previous culture. There are now commercially available matrices (e.g. matrigel, pronectin, cell-tak).

Feeder Layers

Some of the tissue cultures require the support of metabolic products from living cells e.g. mouse embryo fibroblasts where the growing fibroblasts release certain products which when fed to new cell enhance their growth.

Alternative Substrates as Substitutes of Culture Vessels

In recent years, certain alternatives for culture vessels have been developed. The important alternative artificial substrates are micro-carriers and metallic substrates.

Micro-carriers

They are in bead form and are made up of collagen, gelatin, polyacrylamide and polystyrene. Micro-carriers are mostly used for the propagation of anchorage-dependent cells in suspension.

Metallic Substrates

Certain types of cells could be successfully grown on some metallic surfaces or even on the stainless steel discs. For instance, fibroblasts were grown on palladium.

Use of Non-Adhesive Substrates in Tissue Culture

The growth of anchorage-independent cells can be carried out by plating cells on non-adhesive substrates like agar, agarose and methylcellulose. In this condition while the cell growth occurs, the parent and daughter cells get immobilized and form a colony, although they are non-adhesive.

7.4 ANIMAL CELL CULTURE

Animal cell culture is prone to several routes of contamination and they include various glasswares, equipments, reagents, cell lines and poor techniques. The routes of contamination are also associated with the laboratory environment and operating techiques.

Types of Microbial Contamination

Several species of bacteria, yeasts, fungi, molds and mycoplasms, besides viruses are responsible for contamination. Major problems of contamination are linked to the repeated recurrence of a single species. Despite utmost care, no laboratory can claim to be totally free from contamination. It is necessary to continuously monitor for contamination and eliminate the same at the earliest.

The major routes of contamination in a tissue culture laboratory include equipment and facilities. Equipments include Laminar-flow hoods, dry incubators, CO_2 incubators, Humidified incubators and other instruments. Glassware and reagents include pipettes, screw caps, culture glasses, media bottles and biological materials include infected tissue samples and cell lines. Operating techniques include contamination routes such as operator hands, hair, clothing, breathing, work spaces, pipetting and dispensing and operating manipulators.

Aseptic Conditions

Maintenance of proper aseptic conditions is necessary to eliminate various contaminants (due to different microorganisms and viruses). The following measures are suggested for minimizing contamination and maintenance of aseptic conditions:

- Strict adherence to standard sterile techniques and code of practices
- Checking of reagents and media for sterility before use
- Checking of cultures visually and microscopes (phase contrast) every time they are used
- Use of media and separate bottles for each cell line is advised
- Maintenance of clean and tidy conditions at work places

Sterilization

The sterilization procedures are designed to kill the microorganisms, besides destroying the spores. There are three major devices for sterilization:

Dry heat
Moist heat (autoclave)
Filters

Sterilization by dry heat: This is carried out at a minimum temperature of 160^{0}C for one hour.

Sterilization by moist heat: Certain fluids and perishable items can be sterilized in an autoclave at 121^{0}C for 15-20 minutes. For effective moist heat sterilization, it is necessary that the steam penetrates to all parts of the sterilizing materials.

Sterilization by filters: The use of filters for sterilization of liquids often becomes necessary since the constituents of these liquids may get destroyed at higher temperatures (dry heat or moist heat).

Sterile filtration is a novel technique for heat labile solutions: The size of micropores of the filters is 0.1-0.2 μm. Filters made from several materials such as nylon, cellulose acetate, cellulose nitrate, polycarbonate, polyethersulfone (PES) and ceramics are in use. The filters are made in different designs namely disc filters cartridges and hollow fibre. In fact, many commercial companies (e.g. Millipore, Durapore) supply *reusable* and *disposable filters,* designed for different purposes of sterilization.

Items such as glass slides, pipettes, ampoules (glass), Pasteur pipettes, instruments and test tubes can be sterilized using dry heat. Autoclave can be used to sterilize ampoules (plastic), apparatus with silicone tubing, filters (reusable), glass bottles with screw, magnetic stirrer bases, screw caps, stoppers and rubber silicone, are required.

Nutrients and media such as salt solutions, glucose (20%), agar, bacto-peptone, glycerol, lactalbumin hydrolysate, phenol red, tryptose, HEPES, EDTA and water. Filter sterilization is carried out to sterilize serum, amino acids, vitamin, antibiotics, bovine serum albumin, collagenase, glutamine, drugs, NaOH, trypsin and transferrin are required.

Applications of Animal Cell Culture

Due to the widespread concern on the extensive use of animals for laboratory experiments being not morally and ethically justifiable, some researchers prefer to utilize animal cell culture wherever possible for various studies.

The major applications of laboratory animal cell cultures are as follows:

- Studies on in traceable activity e.g. cell cycle and differentiation, metabolism
- Elucidation of in traceable flux e.g. hormonal receptors and signal transduction
- Studies related to cell to cell interactions e.g. cell adhesion and motility, and metabolic cooperation
- Evaluation of environmental interactions e.g. cytotoxicity, mutagenesis
- Intracellular activity : Studies related to cell cycle and differentiation, transcription, translation, energy metabolism and drug metabolism
- Intracellular flux : Studies involving hormonal receptors, metabolites, signal transduction and membrane trafficking
- Cell to cell interaction : Studies dealing with cell adhesion and motility matrix interaction, morphogenesis, paracine control and metabolic cooperation

- Environmental Interaction : Studies related to drug actions, infections, cytotoxicity, mutagenesis and carcinogenesis
- Genetics : Studies dealing with genetic analysis, transfection, transformation, immortalization and senescence
- Cell products: Wide range of applications of the cellular products formed. E.g. Vaccines, hormones, interferons etc
- Laboratory production of medical/pharmaceutical compounds for wide range of applications e.g. vaccines, interferons and hormones.

There are, however, several limitations on the use of animal cell cultures. This is mostly due to the differences that exist between the *in vivo* and *in vitro* systems, and the validity of the studies conducted in the laboratory.

Products of Animal Cell Cultures

Production of vaccines

Polio Vaccines - Poliomyelitis prophylaxis

Measles vaccine - Measles prophylaxis

Rabies vaccine - Rabies prophylaxis

Malaria vaccine - Malaria prophylaxis

HIV vaccine - AIDS prophylaxis and treatment

Production of therapeutics

1. Plasminogen Activator
   Tissue type
   Urokinase-type
   Recombinant
   Applications: Acute myocardial infarction, Pulmonary embolism, deep vein thrombosis and acute stroke
2. Interferon's:
   Interferon-Anticancer
   Interferon-Immunomodulator Anticancer
   Interferon-y: Antiviral
3. Blood clotting factors
   Factors VII, VIII, IX and X
   Applications: Hemophilia as blood clotting agents
4. Hormones
   Human growth hormone: Growth retardation in children
   Somatotropin: Growth retardation in children
   Follicle Stimulating hormone: Treatment of infertility
   Human chorionic gonadotropin: Treatment of infertility
5. Monoclonal antibodies
   Anti-lipopolysaccharide: Treatment of sepsis
   Human B-cell: Treatment of B-cell

Lymohimas: Lymphoma
Anti-fibrin 99: Diagnosis of blood clot by imaging
Tcm-Fab (breast): Diagnosis of breast cancer

6. Erythropoietin
   Interleukin – 2: Antianaemic agent
   Tumor necrosis factor: Anticancer, HIV treatment
   Granulocyle stimulating factor: Anticancer
   Carcinoembryonic antigen: Anticancer and Diagnosis and monitoring of cancer patients

Animal Cells Modification and their Applications

It is now possible to genetically modify the animal (mammalian) cells to introduce the genes needed for the production of a specific protein or to improve the characteristics of a cell line. The following methods are used to introduce foreign DNA into mammalian cells:

Electroporation

Lipofection

Microinjection

Fusion of mammalian cells with bacteria or viruses

As the foreign DNA gets integrated into the mammalian cellular genome, the gene expresses to produce the desired protein. It is however, necessary to select the best producing recombinant cells by conventional methods using selectable marker genes. The following selectable markers are used for choosing the transfected cells:

Vital thymidine kinase

Bacterial dihydrofolate reductase

Bacterial neomycin phosphotransferase

It is possible to overproduce several proteins in mammalian cells through genetic manipulation e.g. tissue plasminogen activator erythropoietin, interleukin-2, interferon-?, clotting factors VIII and IX, and tumor necrosis factors. The recombinant mammalian cells are conveniently used for the production of monoclonal antibodies which have wide range of applications

Category of Risks & Safety Issues

- Maintenance risks: Age and condition of various equipments and leakage of disposals.
- Personnel risks: Inadequate training, lack of concentration and interest
- Physical risks: Electric shocks, fire and intense cold
- Chemical risks: Toxicity due to poisons, carcinogens, mutagens, irritants and allergies
- Biohazards: Pathogenic organisms, viruses, genetic manipulations, culture cells and DNA (quality and quantity)
- Radioisotope risks: Energy emission and its penetration and ionization

Some of the general safety regulations to minimize the risks associated with cell culture laboratories are as follows:

- Periodical meetings and discussions of local safety committees

- Regular monitoring of the laboratories
- Periodical training of the personnel through seminars and workshops
- Print and make the standard operating procedures (SOPs) available to all staff
- Good record keeping
- Limited access to the laboratory (only for the trained personnel and selected visitors)
- Appropriate waste disposal system for biohazards radioactive wastes, toxins and corrosives

Biohazards

The accidents or the risks associated with the biological materials are regarded as biohazards or biological hazards. There are two main systems that contribute to the occurrence of biohazards:

1. Direct sources of the biological materials
2. Processes or operations involved in their handling

Control of Biohazards

Biohazards can be controlled to a large extent by strict adherence to the regulatory guidelines and maintenance programmes. Some important aspects are listed hereunder:

- Microbiological safety cabinet or biohazard hod with pathogen trap filters
- Vertical laminar-flow hood (instead of horizontal laminar-flow hood), which minimizes the direct exposure of the operator to the samples/processes.
- Pathogen containing samples are treated in separate rooms with separate facilities (centrifuge, incubator, cell counting etc.)
- Sterilization of all wastes, solid glassware etc. and proper disposal
- Facilities for change of clothing while entering and leaving the rooms
- Strict adherence to the access of designated personnel to the culture rooms

Sources of Biohazards

Biological materials

- Tissue samples and cultures with pathogens
- Human cells infected with viruses (including retroviruses)
- Cells subjected to various genetic manipulations

Culture Media for Animal Cells

The selection of an appropriate growth medium for the *in vitro* cultivation of cells is an important and essential step. The mammalian cells of an organ in the body receive nutrients from blood circulation. For culturing these cells *in vitro*, it is expected that they should be provided with the components similar to those present in blood. In general, the choice of medium mostly depends on the type of cells to be cultured, and the purpose of the culture (growth, differentiation and production of desired products). The culture media may be natural or artificial.

Natural Media

In the early years, the natural media obtained from various biological sources were used.

Body fluids

Plasma, serum, lymph, amniotic fluid, acidic and pleural fluids, aqueous humor from eyes and insect hemolymph were in common use. These fluids were tested for sterility and toxicity before their utility.

Tissue extracts

Among the tissue extracts *chick embryo* extract was the most commonly employed. The extracts of liver, spleen, bone marrow and leucocytes were also used as culture media.

Artificial Media

The artificial media (containing partly defined components) have been in use for cell culture since 1950. The minimal criteria needed for choosing a medium for animal cell cultures are listed below.

- The medium should provide all the nutrients to the cells
- Maintain the physiological pH around 7.0 with adequate buffering
- The medium must be sterile and isotonic to the cells

The basis for the cell culture media is the **balanced salt solution** which is originally **used to create a physiological pH and osmolarity** required to maintain cells *in vitro*. For promoting growth and proliferation of cells, various constituents (glucose, amino acids, vitamins, growth factors, antibiotics etc.) are added and several media have been developed. Addition of serum to various media is a common practice. However, some workers in recent years have started using serum-free media.

The physicochemical properties of media required for tissue cultures are briefly described. This is followed by a brief account on balanced salt solutions, commonly used culture media and the serum-free media.

Physicochemical Properties of Culture Media

The culture medium is expected to possess certain physicochemical properties (pH, O_2, Co_2, buffering osmolarity, viscosity, temperature etc) to support good growth and proliferation of the cultural cells.

pH

Most of the cells can grow at a pH in the range of 7.0-7.4, although there are slight variations depending on the type of cells (i.e. cell lines). The indicator phenol red is most commonly used for visible detection of pH of the media. Its coloration at the different pH is shown below:

pH 7.4 - Red
pH 7.0 - Orange
pH 6.5 - Yellow
pH 7.8 - Purple

Carbon dioxide in the medium is in a dissolved state, the concentration of which depends on the atmospheric CO_2, and temperature. CO_2 in the medium exists as carbonic acid (H_2CO_3), and bicarbonate (HC_3) and H^+ ions as shown below.

$$CO_2 + H_2O \text{——} H_2CO_3 \text{——} H^+ + HCO_3$$

As is evident from the above equation, the concentrations of CO_2, HCO_3 and pH are interrelated. By increasing the atmospheric CO_2 the pH will be reduced, making the medium acidic. Addition of sodium bicarbonate (as a component of bicarbonate buffer) neutralizes bicarbonate ions.

$$NaHCO_3 \longrightarrow Na^+ + HCO_3$$

In fact the commercially available media contain a recommended concentration of bicarbonate, and CO_2 for the required pH. In recent years HEPES (hydroxyethyl piperazine 2 – solution acid) buffer, which is more efficient than bicarbonate buffer is being used in the culture media. However, bicarbonate buffer is preferred by most workers because of the low cost, less toxicity and nutritional benefit to the medium. This is in contrast to HEPES, which is expensive, besides being toxic to the cells.

The presence of pyruvate in the medium results in the increased endogenous production of CO_2 by the cells. This is advantageous since the dependence on the exogenous supply of CO_2 and HCO_3 will be less. In such a case, the buffering can be achieved by high concentration of amino acids.

Oxygen

A great majority of cells *in vitro* are dependent on the O_2 supply for respiration. This in fact made possible by a continuous supply of O_2 to the tissues by hemoglobin.

The cultured cells mostly rely on the dissolved O_2 in the medium, which may be toxic at high concentration due to the generation of free radicals. Therefore, it is absolutely necessary to supply adequate quantities of O_2 so that the cellular requirements are met, avoiding toxic effects. Free-radical scavengers (glutathione and mercaptoethanol) are also added to nullify the toxicity. Addition of selenium to the medium is also advocated to reduce toxicity. This is because selenium is a cofactor for the synthesis of glutathione.

In general, the glycolysis, occurring in cultured cells is more anaerobic when compared to *in vivo* cells. Since the depth of the culture medium influences the rate of diffusion, it is advisable to keep the depth of medium in the range 2-5 mm.

Temperature

In general, the optimal temperature for a given cell culture is dependent on the body temperature of the organism, serving as the source of the cells. Accordingly, for cells obtained from humans and warm-blooded animals, the optimal temperature is 37°C. *In vitro* cells cannot tolerate higher temperature and most of them die if the temperature goes beyond 40°C. It is therefore absolutely necessary to maintain a constant temperature (± 0.5°C) for reproducible results.

If the cells are obtained from birds, the optimal temperature is slightly higher (38.5°C) for culturing. For cold-blooded animals (poikilotherms) that do not regulate their body heat (e.g. cold-water fish), the culture temperature may be in the range of 15-25°C. Besides directly influencing growth of cells temperature also affects the solubility of CO_2 i.e. higher temperature enhances solubility.

Osmolality

In general, the osmolality for most of the cultured cells (from different organisms) is in the range of 260-320 mosm/kg. This is comparable to the osmolality of human plasma (290mosm/kg). Once an osmolality is selected for a culture medium, it should be maintained at that level (with an allowance of ±10 mosm/kg). Whenever there is an addition of acids, bases, drugs etc., to the medium the osmolality gets affected. The instrument osmometer is employed for measuring osmolalities in the laboratory.

Balanced Salt Solutions

The balanced salt solutions (BSS) are primarily composed of inorganic salts. Sometimes sodium bicarbonate, glucose and HEPES buffer may also be added to BSS. Phenol red serves as a pH indicator. The important functions of balanced salt solutions are listed hereunder:

- Supply essential inorganic ions
- Provide the requisite pH
- Maintain the desired osmolality
- Supply energy from glucose

In fact, balanced salt solutions form the basis for the preparation of complete media with the requisite additions. Further, BSS is also useful for a short period (up to 4 hours) incubation of cells.

Complete Culture Media

In the early years, balanced salt solutions were supplemented with various nutrients (amino acids, vitamins, serums, etc.) to promote proliferation of cells in culture. Eagle, a pioneer in media formulation, determined (during 1950-60) the nutrient requirements for mammalian cell cultures. Many developments in media preparation have occurred since then. There are many types of media now available for different types of cultures. Some of them are:

EMEM – Eagle's minimal essential medium

DMEM – Dulbecco's modification of Eagle's medium

GMEM – Glasgow's modification of Eagle's medium

RPMI 1630 and RPMI 1640 – Media from Rosewell Park Memorial Institute

The complete media, in general, contains a large number of components such as amino acids, vitamins, salts, glucose, other organic supplements, growth factors and hormones, and antibiotics, besides serum. Depending on the medium, the quality and quantity of the ingredients vary.

Amino acids

All the essential amino acids (which cannot be synthesized by the cells) have to be added to the medium. In addition, even the ***non-essential amino acids*** (that can be synthesized by the cells) are also usually added to avoid any limitation of their cellular synthesis. Among the non-essential amino acids, glutamine and / or glutamate are frequently added in good quantities to the media since these amino acids serve as good sources of energy and carbon.

Vitamins

The quality and quantity of vitamins depends on the medium. For instance, Eagle's MEM contains only water-soluble vitamins (e.g., B-complex, choline, and inositol). The other vitamins are obtained from the serum added. The medium M 199 contains all the fat-soluble vitamins (A, D, E and K) also. In general, **for the media without serum, more vitamins in higher concentration are required.**

Salts

The salts present in the various media are basically those found in ***balanced salt solutions*** (Eagle's BSS and Hank's BSS). The salts contribute to cations (Na^+, K^+, Mg^{2+}, Ca^{2+} etc.) and anions (Cl^-, $HCO^-/3$, $SO^2/$

$_4^-$, $PO^3/_4^-$), and are mainly responsible for the maintenance of osmolality. There are some other important functions of certain ions contributed by the salts:

- Ca^{2+} ions are required for cell adhesion, in signal transduction, besides their involvement in cell proliferation and differentiation.
- Na^+, K^+ and Cl^- ions regulate membrane potential.
- $PO^3/_4^-$, $SO^2/_4^-$ and $HCO^-/_3$ ions are involved in the maintenance of intracellular charge; besides serving as precursors for the production of certain important compounds e.g. $PO^3/_4^-$ is required for ATP synthesis.

Glucose

Majority of culture media contain glucose, which ***serves as an important source of energy***. Glucose is degraded in glycolysis to form pyruvate/lactate. These compounds on their further metabolism enter citric acid cycle and get oxidized to CO_2. However, experimental evidence indicates that the contribution of glucose for the operation of citric acid cycle is very low in vitro (in culture cells) compared to *in vivo* situation. ***Glutamine*** rather than glucose ***supplies carbon for the operation of citric acid cycle***. And for this reason, the cultured cells require very high content of glutamine.

Hormones and growth factors

For the media with serum, addition of hormones and growth factors is usually not required. They are frequently added to serum-free media. Hydrocortisone promotes cell attachment, while insulin facilities glucose uptake by cells. Growth hormone, in association with somatormedins (IGFs), promotes cell proliferation. There are growth factors in the serum that stimulate the proliferation of cells in the culture.

- Platelet-derived growth factor (PDGF)
- Fibroblast growth factor (FGF)
- Epidermal growth factor (EGF)
- Vascular endothelial growth factor (VEGF)
- Insulin-like growth factors (IGF-1, IGF-2)

Organic supplements

Several additional organic compounds are usually added to the media to support cultures. These include certain proteins, peptides, lipids, nucleosides and citric acid cycle intermediates. For serum-free media, supplementation with these compounds is very useful.

Antibiotics

In the early years, culture media invariably contained antibiotics. The most commonly used antibiotics were ampicillin, penicillin, gentamycin, erythromycin, kanamycin, neomycin and tetracycline. Antibiotics were added to reduce contamination. However, with improved aseptic conditions in the present day tissue culture laboratories, the addition of antibiotics is not required. In fact, the use of antibiotics is associated with several disadvantages, namely:

- Possibility of developing antibiotic-resistant cells in culture
- May cause antimetabollic effects and hamper proliferation
- Possibility of hiding several infections temporarily
- May encourage poor aseptic conditions

The present recommendation is that for the routine culture of cells, antibiotics should not be added. However, they may be used for the development of primary cultures.

Serum

Serum is a natural biological fluid and is rich in various components to support cell proliferation. The major constituents found in different types of sera are listed in Table 34.3. The most commonly used sera are calf serum (CS), fetal bovine serum (FBS), horse serum and human serum. While using human serum, it must be screened for vital diseases (hepatitis B, HIV). Approximately 5-20% (v/v) of serum is mostly used for supplementing several media. Some of the important features of the serum constituents are briefly described.

Major Constituents of Serum

Proteins

Albumin

Globulins

Fetuin

Fibronectin

Transferrin

Protease inhibitors

(α_1-antitrypsin)

Amino acids

Almost all the 20

Lipids

Cholesterol

Phospholipids

Fatty acids

Carbohydrates

Glucose

Hexosamine

Other organic compounds

Lactic acid

Pyruvic acid

Polyamines

Urea

Vitamins

Vitamin A

Folic acid

Growth factors

Epidermal growth factor
Platelet-derived growth factor
Fibroblast growth factor

Hormones

Hydrocortisome
Thyroxine
Triodothyromine
Insulin

Inorganics

Calcium
Sodium
Potassium
Chlorides
Iron
Phosphates
Zinc
Selenium

Proteins

The *in vitro* functions of serum protein are not very clear. Some of them are involved in promoting cell attachment and growth e.g. fetuin and fibronectin. Proteins ***increase the viscosity*** of the culture medium, besides ***contributing to buffering action***.

Nutrients and metabolites

Serum contains several amino acids, glucose, phospholipids, fatty acids, nucleosides and metabolic intermediates (pyruvic acid, lactic acid etc.). These constituents do contribute to some extent for the nutritional requirements of cells. This may however, be insignificant in complex media with well supplement nutrients.

Inhibitors

Serum may also contain cellular growth inhibiting factors. Majority of them are artifacts e.g. bacterial toxins, antibodies. The natural serum also contains a physiological growth inhibitor namely transforming growth factor β (TGF-β). Most of these growth inhibitory factors may be removed by heat inactivation (at 56°C for 30 minutes).

Selection of Medium and Serum

As already stated, there are many types of media for cell culture. The selection of a particular medium is based on the cell line and the purpose of culturing. For instance, for chick embryo fibroblasts and Heal cells, EMEM is used. The medium DMEM can be used for the cultivation of neurons. The selection of serum is also based on the type of cells being cultured. The following criteria are taken into consideration while choosing serum.

- Batch to batch variations
- Quality control
- Efficiency to promote growth and preservation of cells
- Sterility
- Heat inactivation

Serum –Free Media

Addition of serum to the culture media has been an age-old practice. However, in recent years, certain serum-free media have been developed. It is worthwhile to know the disadvantages associated with the use of serum, and the advantages and disadvantages of serum-free media. The disadvantages of serum in media are:

The variable composition: There is no uniformity in the composition of the serum. It is highly variable (source, batch, season, collection method, processing). Such differences in the composition significantly influence the cells in culture.

Quality control: To maintain a uniform quality of the serum, special tests have to be performed with each batch of serum, before its use.

Table 1 List of Cell Lines and Media/Serum

Cells or cell line	*Medium*	*Serum*
Chick embryo Fibroblasts	EMEM	CS
Chinese hamster Ovary (CHO)	EMEM, Ham's F12	CS
HeLa cells	EMEM	CS
Human leukemia	RPMI1640	FB
Mouse leukemia	Fischer's medium RPMI 1640	FB, HoS
Neurons	DMEM	FB
Mammary epithelium	RPMI 1640, DMEM	FB
Hematopoietic cells	RPMI 1640, Fischer's medium	FB
Skeletal muscle	DMEM F 12	FB, HoS
Glial cells	MEM, F 12 DMEM	FB
3T3 cells	MEM, DMEM	CS

(EMEM-Eagle's minimal essential RPMI 1640-Medium from to Rosewell Park Memorial Institute; DMEM-Dulbecco's modification of Eagle's medium CS-Calf serum; FB-Fetal bovine serum, HoS-Horse serum)

Development of Serum-free Media

While designing serum-free media, it is desirable to identify the various serum constituent and their quantities. The most important constituents of natural serum with reference to their use in cell cultures may be categorized as follows:

- Growth regulatory factors e.g. PDGF, TGF-β
- Cell adhesion factors e.g. vitamins
- Essential nutrients e.g. vitamins, metabolites, minerals, fatty acids
- Hormones e.g. insulin, hydrocortisone

For replacing the serum and development of serum-free media, several constituents should supplement the media. Some highlights are given below:

Advantages and Disadvantages of Serum-free Media

Advantages

The limitations associated with the use of serum in the media (described above) are eliminated in the serum-free media.

Selection of media with defined composition The main advantage of serum-free medium is to control growth of the cells as desired, with a well-defined medium. This is in contrast to the use of serum wherein the growth frequently proceeds in an uncontrolled fashion.

Regulation of differentiation It is possible to use a factor or a set of factors to achieve differentiation of cells with the desired and specialized functions.

Disadvantages

Slow cell proliferation Most of the serum-free media are not as efficient as serum added media in the growth promotion of cells.

Need for multiple media A large number of serum free media need to be developed for different cell lines. This may create some practical difficulties in a laboratory simultaneously handling several cell lines. Another limitation of serum-free medium is that a given medium may not be able to support the different stages of development even for a given cell line. Hence sometimes, separate media may be required even for the same cell line.

Purity of reagents The active serum possesses some amount of protective and detoxifying machinery that can offer a cleansing effect on the apparatus and reagents. And therefore, in the absence of serum, pure grade reagents and completely sterile apparatus should be used.

Availability and cost In general, the serum-free media are costlier than the serum added media. This is mainly due to the fact that many of the pure chemicals added to the serum-free media are themselves expensive. Further, the availability of serum-free media is also another limitation.

Table 2 List of Commercially Available Cell Lines along with Serum-free Media

Cell line	*Medium*
Chick embryo fibroblasts	MCDB 202
Chinese hamster ovary (CHO)	MCDB 402
Human lung fibroblasts	MCDB 110
Human vascular endothelium	MCDB 131
Mammary epithelium	MCDB 170
Prostatic epithelium	WAJC 404
Bronchial epithelium	LHC 9
Fibroblasts	MCDB 202
3T3 cells	MCDB 402

Characteristics of Cultured Cells

Some of the important distinguishing properties of cultured cells are given below:

1. Cells which do not normally proliferate *in vivo* can be grown and proliferated in culture.
2. Cell to cell interactions in the cultural cells are very low.
3. The three dimensional architecture of the *in vivo* cells is not found in cultured cells.
4. The hormonal and nutritional influence on the cultured cells differs from that on the in vivo cells.
5. Cultured cells cannot perform differentiated and specialized functions.
6. The environment of the cultured cells favors proliferations and spreading of unspecialized cells.

Environmental Influence on Cultural Cells

The environmental factors strongly influence the cells in culture. The major routes through which environmental influence occurs are listed below:

- The nature of the substrate or phase in which cells grow. For monolayer cultures, the substrate is a solid (e.g. plastics) while for suspension cultures, it is liquid.
- Composition of the medium used for culture nutrients and physicochemical properties
- Addition of hormones and growth factors
- Composition of the gas phase
- Temperature of culture incubation

The biological and other aspects of cultural cells with special reference to the following parameters are briefly described:

1. Cell adhesion
2. Cell proliferation
3. Cell differentiation
4. Metabolism of cultured cells
5. Initiation of cell culture
6. Evolution and development of cell lines

Cell Adhesion

Most of the cells obtained from solid tissues grow as adherent monolayers in cultures. The cells derived from tissue aggregation or subculture, attach to the substrate and then start proliferating. In the early days of culture techniques, slightly negative charged glasses were used as substrates. In recent years, plastics such as polystyrene, after treatment with electric ion discharge, are in use.

The cells adhesion occurs through cell surface receptors for the molecules in the extracellular matrix. It appears that the cells secrete matrix proteins, which spread on the substrate. Then the cells bind to matrix through receptors. It is a common observation that the substrates (glass or plastic) with previous cell culture are conditioned to provide better surface area for adhesion.

Cell adhesion molecules

Three groups of proteins collectively referred to as cell adhesion molecules (CAMs) are involved in the cell-cell adhesion and cell-substrate adhesion.

Cell-cell adhesion molecules

These proteins are primarily involved in cell-to-cell interaction between the homologous cells. CAMs are of two types-calcium-dependent ones (cadherins) and calcium-independent CAMs.

Integrins

These molecules mediate the cell substrate interactions. Integrins possess receptors for matrix molecules such as fibronectin and collagen.

Proteoglycans

These are low affinity transmembrane receptors. Proteoglycans can bind to matrix collagen and growth factors. Cell adhesion molecules are attached to the cytoskeletons of the cultured cells.

Cell Proliferation

Proliferation of cultured cells occurs through the ***cell cycle***, which has four distinct phases

M phase: In this phase (M-mitosis), the two chromatids, which constitute the chromosomes, segregate to daughter cells.

G_1 phase: This gap 1 phase is highly susceptible to various control processes that determine whether cell should proceed towards DNA synthesis, re-enter the cycle or take the course towards differentiation.

S phase: This phase is characterized by DNA synthesis wherein DNA replication occurs.

G_2 phase: This is gap 2 phase that prepares the cell for reentry into mitosis.

The integrity of the DNA, its repair or entry into apoptosis (programmed cell death) if repair is not possible is determined by two check points-at the beginning of DNA synthesis and in G_2 phase.

Control of cell proliferation

For the cells in culture, the environment signals regulate the cell cycle and thereby the cell proliferation. Low density of the cells in a medium coupled with the presence certain growth factors (e.g. epidermal growth factor, platelet-derived growth factor) allows the cells to enter the cell cycle. On the other hand, high cell density and crowding of cell inhibits the cell cycle and thereby proliferation.

Besides the influence of the environment factors, certain intracellular factors also regulate the cell cycle. For instance, cyclins promote while p53 and Rb gene products inhibit cell cycle.

Cell Differentiation

The various cell culture conditions favor maximum cell proliferation and propagation of cell lines. Among the factors that promote cell proliferation the following are important.

- Low cell density
- Low Ca^{2+} concentration
- Presence of growth factors

For the process of cell differentiations to occur, the proliferation of cells has to be severely limited or completely abolished. Cell differentiation can be promoted (or induced) by the following factors.

- High cell density

- High Ca^{2+} concentration
- Presence of differentiation inducers (e.g. hydro cortisone, nerve growth factor)

As is evident from the above, different and almost opposing conditions are required for the cell proliferation and differentiation. Therefore if cell differentiation is required two distinct sets of conditions are necessary:

1. To optimize cell proliferation
2. To optimize cell differentiation

Maintenance of Differentiation

The cells retain their native and original functions for long when the dimensional structures are retained. This is possible with organ cultures and however, organ cultures cannot be propagated. Researchers are experimenting to create three-dimensional structures by perfusing monolayer cultures. Further *in vitro* culturing of cells on or in special matrices (e.g., cellulose, collagen gel, matrix of glycoproteins) also results in cell's three-dimensional structures.

Dedifferentiation

Dedifferentiation refers to the irreversible loss of specialized properties of cells when they are cultured *in vitro*. This happens when the differentiated *in vitro* cells lose their properties. Dedifferentiation implies an irreversible loss of specialized properties of the cells. On the other hand, deadaptation refers to the reinduction of specialized properties of the cells by creating appropriate conditions.

In the *in vivo* situation, a small group of stem cells give rise to progenitor cells that are capable of producing differentiated cell pool. On the other hand, in the vitro culture system, progenitor cells are predominantly produced which go on proliferating. Very few of the newly formed cells can form differentiated cells. The net result is a blocked differentiation.

Metabolism of Cultured Cells

The metabolism of mammalian cultured cells with special reference to energy use glucose or glutamine as the source of energy. These two compounds also generate important anabolic precursors.

As glucose gets degraded by glycolysis, lactate is mainly produced. This is because oxygen is in limited supply in the normal culture conditions (i.e., atmospheric oxygen and a submerged culture) creating an anaerobic situation. Lactate, secreted into the medium accumulates. Some amount of pyruvate produced in glycolysis gets oxidized through Krebs cycle. A small fraction of glucose (4-9%) enters pentose phosphate pathways to supply ribose 5-phosphate and reducing equivalent (NADPH) for biosynthetic pathways, e.g., synthesis of nucleotides.

Glutamine is an important source of energy fort he cultured cells. By the action of he enzyme glutaminase, glutamine undergoes deamination to produce glutamate and ammonium ions. Glutamate, on transamination (or oxidative deamination) forms a-ketoglutarate which enters the Krebs cycle. Pyruvate predominantly participates in transamination reaction to produce alanine, which is easily excreted into the medium. In the rapidly growing cultured cells, transamination reaction is a dominant route of glutamine metabolism.

Deamination of glutamine release free ammonium ion, which are toxic to the cultured cells, limiting their growth. In recent years, dipeptides glutamyl-alanine or glutamyl-glycine is being used to minimize the production of ammonia. Further, these dipeptides are more stable in the medium.

An already stated α-ketoglutarate obtained from glutamine (via glutamate) enters the Kreb's cycle and gets oxidized to carbon dioxide and water. For proper operation of Kreb's cycle balancing of the intermediates of the cycle is required. Two metabolites of Kreb's cycle namely malate and oxaloacetate leave the cycle and get converted respectively to pyruvate and phosphoenol pyruvate. The latter two compounds can reenter the Kreb's cycle in the form of acetyl CoA. Thus, the continuity of Kreb's cycle is maintained. Glucose as well as glutamine gets metabolized by the cultured cells to supply energy in the form of ATP.

Initiation of Cell Culture

The cell culture can be initiated by the cells derived from a tissue through enzymatic or mechanical treatments. Primary culture is a selective process that finally results in a relatively uniform cell line. The selection occurs by virtue of the capacity of the cells to survive as monolayer cultures (by adhering to substrates) or as suspension cultures.

Among the cultured cells, some cells can grow and proliferate while some are unable to survive under the culture environment. The cells continue to grow in monolayer cultures, till the availability of the substrate is occupied.

The term confluence is used when the cultured cells make close contact with one another by fully utilizing the available growth area. For certain cells, which are sensitive to growth limitations due to density, the cells stop growing once confluence is reached. However, the transformed cells are insensitive to confluence and continue to overgrow.

When the culture becomes confluent, the cells possess the following characters:

1. Closest morphological resemblance to the tissue of origin (i.e. parent tissue)
2. Expression of specialized functions of the cells comparable to that of the native cells

Evolution and Development of Cell Lines

The primary culture grows after the first subculture is referred to as cell line. A given cell line may be propagated by further sub culturing, As the subcultures are repeated, the most rapidly proliferating cells dominate while the non-proliferating or slow proliferating cells will get diluted, and consequently disappear.

Senescence

The genetically determined event of cell division for a limited number of times (i.e. population doublings), followed by their death in a normal tissue is referred to as senescence. However, germ cells and transformed cells are capable of continuously proliferating. In the *in vitro* culture, the transformed cells can give rise to continuous cell lines.

Development of continuous cell lines

Certain alterations in the culture collectively referred to as transformation can give rise to continuous cell lines. Transformation may be spontaneously occurring, chemically or virally-induced. Transformation basically involves an alteration in growth characteristics such as loss of contact inhibition, density limitation of growth and anchorage independence. The term immortalization is frequently used for the acquisition of infinite life span to cultured cells.

Genetic variation

The ability of the cells to grow continuously in cell lines represents genetic variation in the cells. Most often, the deletion or mutation of the p^{53} gene is responsible for continuous proliferation of cells. In the normal cells, the normal p^{53} gene is responsible for the arrest of cell cycle.

Most of the continuous cell lines are aneuploid, possessing chromosome number between diploid and tetraploid value.

Normal cells and continuous cell lines

A great majority of normal cells are not capable of giving rise to continuous cell lines. For instance, normal human fibroblasts go on proliferating for about fifty generations, and then stop dividing. However, they remain viable for about eighteen months. And throughout their life span, fibroblasts remain euploid. Chick fibroblasts also behave in a similar fashion. Epidermal cells and lymphoblast cells are capable of forming continuous cell lines.

Characterization of Cultured Cells

Characterization of cultured cells or cell lines is important for dissemination of cell lines through cell banks, and to establish contacts between research laboratories and commercial companies. Characterization of cell lines with special reference to the following aspects is generally done:

1. Morphology of cells
2. Species of origin of cells
3. Tissue of origin
4. Filament proteins as tissue markers
5. Cell surface antigens as tissue markers

Morphology of cells

A simple and direct identification of the cultured cells can be done by observing their morphological characteristics. However, the morphology has to be viewed with caution since it is largely dependent on the culture environment. For instance, the epithelial cells growing at the center (of the culture) is regular polygonal with clearly defined edges, while those growing at the periphery are irregular and distended (swollen). The composition of the culture medium and the alterations in the substrate also influence the cellular morphology. In a tissue culture laboratory, the terms fibroblastic and epithelial are commonly used to describe the appearance of the cells rather than their origin.

Fibroblastic cells: For these cells, the length is usually more that twice if their width. Fibroblastic cells are bipolar or multipolar in nature.

Epithelial cells: These cells are polygonal in nature with regular dimensions and usually grow in monolayer.

The terms fibroblastic (fibroblast-like) and epithelial (epithelial-like) are in use for the cells that do not specific characters to identify as fibroblastic or epithelial cells.

Species of origin of cells

The identification of the species of cell lines can be done by:

- Chromosomal analysis

- Electrophoresis of isoenzymes
- A combination of both these methods

In recent years chromosomal identification is being done by employing molecular probes.

Tissue of origin

The identification of cell lines is based on two characteristic features:

1. The lineage to which the cells belong
2. The status of the cells i.e. stems cells or precursor cells

Tissue markers for cell line identification Some of the important tissue or lineage markers for cell line identification are briefly described.

Differentiated products as cell markers The cultured cells to complete expression are capable of producing differentiation markers, which serve as cell markers for identification. Some are given below:

- Albumin for hepatocytes
- Melanin for melanocytes
- Hemoglobin for erythroid cells
- Myosin (or tropomyosin) for muscle cells

Enzymes as tissue markers The identification of enzymes in culture cells can be made with reference to the following characters.

- Constitutive enzymes
- Inducible enzymes
- Isoezymes

The commonly used enzyme markers for cell line identification are given in Table 35.1.

Tyrosine aminotransferase is specific for hepatocytes while tyrosinase is for melanocytes. Creatine kinase (MM) in serum serves as a marker for muscle cells, while creatine kinase (BB), and is used for the detection of neurons and neuroendocrine cells.

Table 3 Some Common Enzyme Markers for Cell Line Identification

Enzyme	*Cell type*
Tyrosine aminotraslerase	Hepatocytes
Tysosinase	Melanocytes
Glutamyl synthase	Brain (astroglia)
Creatine kinase (isoenzyme MM)	Muscle cells
Creatine kinase (isoenzyme BB)	Neurons
Non-specific esterase	Macrophages
DOPA-decarboxylase	Neurons
Alkaline phosphatase	Enterocytes, type it pneumocytes
Angiotensin-converting enzyme	Endothellum
Sucrase	Entherocytes
Neuron-specific esterase	Neurons

DOPA-Dihydroxy phenylalanine, MM-two polypeptide subunits of muscle; BB-Two polypeptide subunits of brain.

Table 4 A Selected List of Antibiotics used for the Detection of Cell Types

Antibody	*Cell type*
Cytokeratin	Epithelium
Epithelial membrane antigen	Epithlium
Albumin	Hepatocytes
α-Lactalbumin	Breast epithelium
Carcinoembryonic antigen (CEA)	Colorectal and lung adenicacinoma
Prostate specific antigen (PSA)	Prostatic epithelium
Intracellular cell adhesion molecule (I-CAM)	T-cells and endothelium
α-Fetoprotein	Fetal hepatocytes
Human chorionic Gonadotropin (hCG)	Placental epithelium
Human growth hormone (HGH)	Anterior pituitary
Vimentin	Mesodermal cells
Integrins	All cells
Actin	All cells

Filament proteins as tissue markers

The intermediate filament proteins are very widely used as tissue or lineage markers. For example,

- Astrocytes can be detected by glial fibrillary acidic protein (GFAP)
- Muscle cells can be identified by desmin
- Epithelial and mesothelial cells by cytokeratin

Cell surface antigens as tissue markers

The antigens of the cultured cells are useful for the detection of tissue or cells of origin. In fact, many antibodies have been developed (commercial kits are available) for the identification cell lines (Table 35.2). These antibodies are raised against cell surface antigens or other proteins.

The antibodies raised against secreted antigen α-fetoprotein serve as a marker for the identification of fetal hepatocytes. Antibodies of cell surface antigens namely integrins can be used for the general detection of cell lines.

7.5 TRANSFORMED CELLS

Transformation if the phenomenon of the change in phenotype due to the acquirement of new genetic material. Transformation is associated with promotion of genetic instability. The transformed and cultured cells exhibit alterations in many characters with reference to:

- Growth rate
- Mode of growth
- Longevity
- Tumorigenicity
- Specialized product formation

While characterizing the cell lines, it is necessary to consider the above characters to determine whether the cell line has originated from tumor cells or has undergone transformation in culture.

Identification of Specific Cell Lines

There are many approaches in a culture laboratory to identify specific cell lines, such as:

- Chromosome analysis
- DNA detection
- RNA and protein analysis
- Antigenic markers

Chromosome analysis

The species and sex from which the cell line is derived can be identified by chromosome analysis. Further, it is also possible to distinguish normal and malignant cells by the analysis of chromosomes. It may be noted that the normal cells contain more stable chromosomes. The important techniques employed with regard to chromosome analysis are briefly described below:

Chromosome banding: By this technique, it is possible to identify individual chromosome pairs when there is little morphological difference between them. Chromosome banding can be done using Giemsa staining.

Chromosome count: A direct count of chromosomes can be done per spread between 50-100 spreads. A *camera lucida* attachment or a closed circuit television may be useful.

Chromosome karyotyping: In this technique, the chromosomes are cut, sorted into sequence, and then pasted on to a sheet. The image can be recorded or scanned from the slide. Chromosome karyotyping is time consuming when compared to chromosome counting.

DNA detection

The total quantity of DNA per normal cell is quite constant, and is characteristic to the species of origin e.g. normal cell lines from human, chick and hamster fibroblasts. However, the DNA content varies in the normal cell lines of mouse, and also the cell lines obtained for cancerous tissues. As already stated, most of the transformed cells are aneuploid. DNA analysis is particularly useful for characterization of such cells. Analysis of DNA can be carried out by DNA hybridization and DNA fingerprinting.

DNA hybridization: The popular Southern blotting technique (for details, Refer Chapter 7), can be used to detect unique DNA sequences. Specific molecular probes with radioisotope, fluorescent or luminescent labels can be used for this purpose.

The DNA from the desired cell lines is extracted, cut with restriction endonucleases, subjected to electrophoresis, blotted on to nitrocellulose, and then hybridized with a molecular (labeled) probe, or a set of probes. By this approach, specific sequences of DNA in the cell lines can be detected.

DNA fingerprinting: There are certain regions in the DNA of a cell that are not transcribed. These regions, referred to as satellite DNA, have no known functions, and it is believed that they may provide reservoir for genetic evolution. Satellite DNA regions are considered as regions of hypervariability. These regions may be cut with specific restriction endonucleases, and detected by using cDNA probes.

By using electrophoresis and autoradiography, the patterns of satellite DNA variations can be detected. Such patterns referred as DNA fingerprints are cell line specific. In recent years, the technique of DNA fingerprints has become a very popular and a powerful tool to determine the origin of cell lines.

RNA and protein analysis

The phenotype characteristics of a cell line can be detected by gene expression i.e. identification of RNAs and / or proteins. MRNAs can be identified by Northern blot technique while proteins can be detected by Western blot technique.

Enzyme activities Some of the in vivo enzyme activities are lost when the cells are cultured in vitro. For instance, anginase activity of the liver cells is lost within a few days of culturing. However, certain cell lines express specific enzymes that can be employed for their detection. For example, tyrosine aminotransferase for hepatocytes and glutamyl synthase activity for astroglia in brain. For more examples of enzymes useful in cell line detection refer Table 35.1.

Isozymes: The multiple forms of an enzyme catalyzing the same reaction are referred to an isoenzymes or isozymes. Isoenzymes differ in many physical and chemical properties such as structure, electrophoretic and immunological properties K_m and V_{max} values.

The isoenzymes can be separated by analytical techniques such as electrophoresis and chromatography. Most frequently electrophoresis by employing agarose, cellulose acetate, starch and polyacrylamide is used. The crude enzyme is applied at one point on the electrophoretic medium. As the isoenzymes migrate, they distribute in different bands, which can be detected by staining with suitable chromogenic substrates.

Isoenzymes are characteristic to the species or tissue. Isoenzymes of the following enzymes are commonly used for cell line detection.

- Lactate dehydrogenase
- Malate dehydrogenase
- Glucose 6- phosphate dehydrogenase
- Aspartate aminotransferase
- Peptidase B

Isoenzyme analysis is also useful for the detection of interspecies cross-contamination of cell lines. For instance, contamination of mouse cell line with hamster cell line can be identified by using peptidose B isoenzymes/

Antigenic markers

Cell lines can be characterized by detection of antigenic markers through the use of antibodies. The antigenic markers may be located on the cell surface or secreted by the cells into the culture medium.

Some of the antibodies in common use for the detection of different cell types are given in Table 35.2. See page 430.

Measurement of Growth Parameters of Cultured Cells

Information on the growth state of a given culture is required to:

- Design culture experiments

- Routine maintenance of culture
- Measurement of cell proliferation
- Know the time for subculture
- Determine the culture response to a particular stimulus or toxin

Some of the commonly used terms in relation to the measurement of growth of cultured cells are explained.

Population doubling time (PDT)

The time interval for the cell population to double at the middle of the logarithmic (log) phase.

Cell cycle time or generation time

The interval from one point in the cell division to the same point in the cycle, one division later. Thus cell cycle time is measured from one point in the cell cycle until the same point is reached again.

Confluence

It denotes the culture stage wherein all the available substrate (growth area) is utilized, and the cells are in close contact with each other.

Contact inhibition

Inhibition of cell motility and plasma membrane ruffling when the cells are in complete contact with other adjacent cells. This mostly occurs at confluence state, and results in the cessation of the cell proliferation.

Cell density

The number of cells per ml of the medium

Saturation density

The density of the cells (cells/ml^2 surface area) in the plateau phase

Growth Cycle of Cultured Cells

The growth cycle of cultured cells is conventionally represented by three phases, vis-à-vis, the lag phase, the log (exponential) phase and the plateau phase (Fig. 35.5). The properties of the cultured cells vary in the phases.

The lag phase

The lag phase represents a period of adaptation during which the cell forms the cell surface and extra cellular matrix (lost during trypsinization), attaches to the substrate and spreads out. There is an increased synthesis of certain enzymes (e.g. DNA polymerase) and structural proteins, preparing the cells for proliferation. The production of specialized products disappears which may not reappear until the cell proliferation ceases. The lag phase represents preparative stage of the cells for proliferation following subculture and reseeding.

The log phase

The log phase is characterized by an exponential growth of cells, following the lag phase. The duration of log phase depends on the cells with reference to:

- Seeding density

- Growth rate
- Density after proliferation

During the log phase, the cultured cells are in the most uniform and reproducible state with high viability. This is an ideal time for sampling. The log phase terminated after confluence is reached with an addition of one or two population doublings.

The plateau phase

As the cells reach confluence, the growth rate is much reduced, and the proliferation of cultured cells almost stops. This stage represents plateau or stationary phase, and is characterized by low motility of cells, reduced ruffling of plasma membrane, cells occupying minimum surface area, contact inhibition, and saturation density, depletion of nutrients and growth factors, reduced synthesis of structural proteins and increased formation of specialized products.

The majority of normal cultured cells that form monolayer stop growing as they reach confluence. Some of the cells however, with replenishment of medium continue to grow (at a reduced rate) after confluence, forming multilayer of cells. The transformed cultured cells usually reach a higher cell density compared to the normal cells in the plateau phase.

Plating Efficiency of Cultured Cells

Plating efficiency, representing colony formation at low cell density, is a measure used for analyzing cell proliferation and survival.

Cloning efficiency

When the cells, at low densities, are cultured in the form of single cell suspensions, they grow as discrete colonies. Plating efficiency is calculated as follows.

$$\text{Plating efficiency} = \frac{\text{No. of colonies formed}}{\text{No. of cells seeded}} \times 100$$

The term **cloning efficiency** is used (instead of plating efficiency) when each colony grows from a single cell.

Seeding efficiency

Seeding efficiency representing the survival of cells at higher densities is calculated as follows.

$$\text{Seeding efficiency} = \frac{\text{No. of cells recovered}}{\text{No. of cells seeded}} \times 100$$

7.6 CELL SYNCHRONIZATION

Synchronization literally means to make two or more things happen exactly simultaneously. For instance, two or more watches can be synchronized to show exactly the same time.

The cells at different stages of the cell cycle in a culture can be synchronized so that the cells will be at the same phase. Cell synchrony is required to study the progression of cells through cell cycle. Several laboratory techniques have been developed to achieve cell synchronization. They are broadly categorized into two groups.

- Physical fractionation for cell separation
- Chemical blockade for cell separation

Cell Separation by Physical Means

Physical fractionation or cell separation techniques, based on the following characteristics are in use:

- Cell density
- Cell size
- Affinity of antibodies on cell surface epitopes
- Light scatter or fluorescent emission by labeled cells

The two commonly used techniques namely centrifugal elutriation and fluorescence-activated cell separation are briefly described hereunder.

Centrifugal Elutriation

The physical characteristics namely the cell size and sedimentation velocity are operative in the technique of centrifugal elutriation. Centrifugal elutriator (from Beckman) is an advanced device for increasing the sedimentation rate so that the yield and resolution of cells is better. The cell separation is carried out in a specially designed centrifuge and rotor. The cells in the medium are pumped into the separating chamber while the rotor is turning. Due to centrifugal force, the cell will be pushed to the edges.

As the medium is then pumped through the chamber in such a way that the centripetal flow is equal to the sedimentation rate of cells. Due to difference in the cells (size, density, cell surface configuration), the cells tend to sediment at different rated, and reach equilibrium at different positions in the chamber. The entire operation in the elutriator can be viewed through the port, as the chamber is illuminated by stroboscopic light. At the equilibrium the flow rate can be increased and the cells can be pumped out, and separated in collecting vessels in different fractions. It is possible to carry out separation of cells in a complete medium, so that the cells can be directly cultured after separation.

Fluorescence-activated Cells Sorting

Fluorescence-activated cell sorting is a technique for sorting out the cells based on the differences that can be detected by light scatter (e.g. cell size) or fluorescence emission (by pretreated DNA, RNA, proteins, antigens).

The procedure involves passing of a single stream of cells through a laser beam so that the scattered light from the cells can be detected and recorded. When the cells are pretreated with the fluorescent stain (e.g. chromomycin A for DNA), the fluorescent emission excited by the laser can be detected.

There are two instruments in use based on the principle of fluorescent-activated cell sorting:

1. **Flow cytometer:** This instrument is capable of sorting out cells (from a population) in different phases of the cell cycle based on the measurements of a combination of cell size and DNA fluorescence.
2. **Fluorescent-activated cell sorter (FACS):** In this instrument, the emission signals from the cells are measured, and the cells sorted out into collection tubes.

Comparison between Physical Methods

For separation of a large number of cells, centrifugal elutriator is preferred. On the other hand, fluorescent-activated cell sorting is mostly used to obtain high-grade pure frat ions of cells from small quantities of cells.

Cell Separation by Chemical Blockade

The cells can be separated by blocking metabolic reactions. Two types of metabolic blockades are in use vis-à-vis inhibition of DNA synthesis and nutritional deprivation.

Inhibition of DNA Synthesis

During the S phase of cell cycle, DNA synthesis can be inhibited by using inhibitors such as thymidine, aminopterine, bydroxyurea and cytosine arabinoside. The effects of these inhibitors are variable. The cell cycle is predominantly blocked in S phase that results in viable cells.

Nutritional Deprivation

Elimination of serum or isoleucine from the culture medium for about 24 hours results in the accumulation of cells at G_1 phase. This effect of nutritional can be restored by their addition by which time the cell synchrony occurs.

Some Highlights of Cell Synchronization

- Cell separation by physical methods is more effective than chemical procedures
- Chemical blockade is often toxic to the cells
- Transformed cells cannot be synchronized by nutritional deprivation
- A high degree of cell synchrony (>80%) can be obtained in the first cycle, and in the second cycle it would be <60%. The cell distribution may occur randomly in the third cycle.

7.7 CELL SENESCENCE AND APOPTOSIS

As the cells grow in culture, they become old due to aging, and they cannot proliferate any more. The end of the proliferating life span of cells is referred to as senescence.

Cellular Senescence

The growth of the cells is usually measured as population doublings (PDs). The PDs refer to the number of times the cell population doubles in number during the period of culture and is calculated by the following formula.

$$PD = \frac{\log_{10}(\text{No. of cells harvested}) - \log_{10}(\text{No. of cells seeded})}{\text{Log}_{10}^{2}}$$

The phenomenon of senescence has been mostly studied with human fibroblast cultures. After 30-60 populations' doublings, the culture is mainly composed of senescent fibroblasts. This senescent fibroblast is unable to divide in response to mitotic stimuli. It must be noted that the cells do not appear suddenly, but they gradually accumulate and increase in number during the life span of the culture.

The different parameters used for the measurement of cell growth in cultures are listed below:

- Direct measure of cell number
- Determination of DNA/RNA content
- Estimation of protein/ATP concentration

Measurement of Senescence

The direct measurement of senescent cells is rather difficult. Some of the indirect measures are:

- Loss of metabolic activity
- Lack of labeled precursor (^{3}H-thymidine) incorporation into DNA
- Certain histochemical techniques

Senescence-associated β-galactosidase Activity Assay

There occurs an over expression of the lysosomal enzyme β-galactosidase at senescence. This enzyme elevation is also associated with an increase in the cell size as the cell enters a permanent non-dividing state.

The number of senescent cells in a culture can be measured by senescence-associated β-galactosidase (SA-β) assay. The assay consists of the following stages:

1. Wash the cells and fix them using a fixative (e.g. para formaldehyde), and wash again.
2. Add the staining solution (X-gal powder in dimethylformamide dissolved in buffer) to the fixed cells and incubate.
3. The senescent cells display a dense blue color, which can be counted.

Apoptosis

The process of **programmed cell death (PCD)** is referred to as apoptosis. The cell death may be initiated by a specific stimulus or as a result of several signals received from the external environment.

Apoptosis occurs as a result of inherent cellular mechanisms, which finally lead to self-destruction. The cell activates a series of molecular events that cause an orderly degradation of the cellular constituents with minimal impact on the neighboring tissues.

Reasons for in situ apoptosis

1. For proper development: The formation of fingers and toes of the detus requires the removal of the tissues between them. This is usually carried out by apoptosis.
2. Destruction of cells that pose threat to the integrity of the organism: Programmed cell death is needed to destroy and remove the cells that may otherwise damage the organisms. Some examples are listed:
   - Cells with damaged DNA during the course of embryonic development. If they are not destroyed, they may result in birth defects.
   - Cells of the immune system, after their appropriate immune function, undergo apoptosis. This is needed to prevent autoimmune diseases e.g. rheumatoid arthritis.
   - Cells infected with viruses are destroyed by apoptosis.
3. Cell destruction due to negative signals: There are several negative signals within the cells that promote apoptosis. These include accumulation of free radicals, exposure to UV rays, X-rays and chemotherapeutic drugs.

Mechanism of Apoptosis

The programmed cell death may occur due to three different mechanisms:

1. Apoptosis due to internal signals
2. Apoptosis triggered by external signals e.g. tumor necrosis factor -α (TNF-α), lymphotoxin
3. Apoptosis triggered by reactive oxygen species

Role of caspases in apoptosis

A group of enzymes namely **activated proteases** play a crucial role in the programmed cell death. These proteases are actually **c**ysteinyl **a**sparate **s**pecific **p**rotein**ases** or in short, commonly referred to as **caspases**. There are about ten different types of caspases acting on different substrate ultimately leading to cell death. For instance, capsase I cleaves interleukin 1β.

Inhibition of caspase activities

Since the caspases are closely involved in apoptosis, it is possible to prevent cell death by inhibiting their activities. Certain specific peptides that can inhibit caspases, and thus apoptosis have been identified.

Measurement of Apoptosis

A simple and easy way of detecting dead or dying cells is the direct microscopic observation. The dying cells are rounded with dense bodies, which can be identified under phase contrast microscope. The cells that have undergone apoptosis contain fragmented chromatin, which can be detected by conventional staining techniques. In recent years, more sensitive and reliable techniques have been developed for measuring apoptosis. Some of them are briefly described.

Determination ADP/ATP ratio

Both the growth and apoptosis of cells require ATP. But when there is growth arrest, an elevation of ADP occurs. Thus measuring ADP/ATP ratio will throw light on the dead cells. In fact, some assay systems for measuring ADP/ATP ratios are commercially available.

TUNEL assay

A significant biochemical event for the apoptosis is the activation of endogenous nuclease activity. This enzyme cleaves DNA into fragments with free 3-hydroxyl groups. The newly formed small DNA fragments can be extended by employing the enzyme DNA polymerase. If labeled nucleotides are used for DNA fragment extension, they can be detected.

TUNEL is an abbreviation of TdT-mediated dUTP nick end-labeling assay. TUNEL is very fast and effective for the determination of DNA fragments formed by endogenous nuclease activity. The apoptotic nuclei can be identified by a fluorescent technique using fluorescein isothiocyanate (FITC) and 4, 6-diaminophenylindole.

DNA laddering test

During the course of apoptosis, the genomic DNA is cleaved to mono – and oligonucleosomal DNA fragments. These fragments can be separated by agarose electrophoresis, and detected. The nucleosomal fragments of apoptotic cells give a characteristic ladder pattern of electrophoresis.

Limitations of the test DNA laddering test is not very specific since several cells that have undergo9ne apoptosis may not show DNA laddering. Further, some cells not subjected to apoptosis may also show DNA ladders, for these reasons, DNA laddering test is coupled with some other test for measurement of apoptosis.

Safety Measures in Culture Techniques

Since the culture techniques involve the use of animal or human tissues, it is absolutely necessary to follow several safety measures and medical ethics. In fact, in some countries there are established legislation/norms for selection and use of tissues in cultures. For example, in United Kingdom, Animal Experiments (Scientific Procedures) Act of 1986 is followed.

The handling of human tissues poses several problems that are not usually encountered with animal tissues. While dealing with fetal materials and human biopsies, the consent of the patient and/his or her relatives, besides the consent of local ethical committee is required. Further, taking any tissue (even in minute quantities) from human donors requires the full consent of the donor in a prescribed format. The following issues need to be fully considered while dealing with human tissues.

1. The consent of the patient and/or relatives for using tissues for research purposes
2. Ownership of the cell lines developed and their derivatives
3. Consent for genetic modification of the cell lines
4. Patent rights for any commercial use of cell lines

In the general practice of culture techniques using human tissues, the donor and/or relatives are asked to sign a **disclaimer statement** (in a prescribed proforma before the tissue is taken. By this approach, the legal complications are minimized.

Handling of **human tissues** is associated with a heavy **risk of exposure for various infections**. Therefore, it is absolutely necessary that the human materials are handled in a **biohazard cabinet**. The tissues should be screened for various infections such as hepatitis, tuberculosis, HIV, before their use. Further, the media and apparatus, after their use must be autoclaved or disinfected, so that the spread of infections is drastically reduced.

Primary Cultured and Cell Lines

Primary culture refers to the starting culture of cells, tissues or organs, taken directly from an organism. Thus, the primary culture is the initial culture before the first subculture. The term cell line is used for the propagation of cultures after the first subculture. Some basic and fundamental aspects of primary culture and cell lines are briefly described.

Primary Cell Culture

As already stated, primary culture broadly involves the culturing techniques carried following the isolation of the cells, but before the first subculture. The primary cultures are usually prepared from large tissue masses. Thus, these cultures may contain a variety of differentiated cells e.g. fibroblasts, lymphocytes, macrophages and epithelial cells.

With the experiences of the personnel working in tissue culture laboratories, the following criteria/characteristics are considered for efficient development of primary cultures:

- Embryonic tissues rather than adult tissues are preferred for primary cultures. This is due to the fact that the embryonic cells can be disaggregated easily and can yield more viable cells, besides rapidly proliferating *in vitro*.
- Quantity of cells used in the primary culture should be higher since their survival rate is substantially lower (when compared to subcultures).

- The tissues should be processed with minimum damage to cells for use in primary culture. Further, the dead cells should be removed.
- Selection of an appropriate medium (preferably a nutrient rich one) is advisable. For the addition of serum, fetal bovine source is preferred rather than calf or horse serum.
- It is necessary to remove the enzymes used for desegregation of cells by centrifugation.

Techniques for Primary Culture

Among the various techniques devised for the primary culture of isolated tissues, three techniques are most commonly used:

1. Mechanical desegregation
2. Enzymatic desegregation
3. Primary explants technique

Mechanical disaggregation

For the disaggregation of soft tissues (e.g. spleen, brain, embryonic liver, soft tumors), mechanical technique is usually employed. This technique basically involves careful chopping or slicing of tissue into pieces and collection of spill out cells. The cells can be collected by two ways.

- Pressing the tissue pieces through a series of sieves with a gradual reduction in the mesh size
- Forcing the tissue fragments through a syringe and needle

Although mechanical disaggregation involves the risk of cell damage, the procedure is less expensive, quick and simple. This technique is particularly useful when the availability of the tissue is in plenty, and the efficiency of the yield is not very crucial. It must however, be noted that the viability of cells obtained from mechanical techniques is much lower than the enzymatic technique.

Enzymatic disaggregation

Enzymatic disaggregation is mostly used when high recovery of cells is required from a tissue. Disaggregation of embryonic tissues is more efficient with higher yield of cells by use of enzymes. This is due to the presence of less fibrous connective tissue and extra cellular matrix. Enzymatic disaggregation can be carried out by using trypsin, collagenase or some other enzymes.

Disaggregation by trypsin The term trypsinization is commonly used for disaggregation of tissues by the enzyme, trypsin. Many workers prefer to use crude trypsin rather than pure trypsin for the following reasons:

- The crude trypsin is more effective due to the presence of other proteases.
- Cells can tolerate crude trypsin better.
- The residual activity of crude trypsin can be easily neutralized by the serum of the culture media (when serum-free media are used, a trypsin inhibitor can be used for neutralization.).

Disaggregation of cells can also be carried out by using pure trypsin, which is less toxic and more specific in its action. The desired tissue is chopped to 2-3 mm pieces and then subjected to disaggregation by trypsin. There are two techniques of trypsinization: warm trypsinization and cold trypsinization.

Warm trypsinization This method is widely used for disaggregation of cells. The chopped tissue is washed with dissection basal salt solution (DBSS), and then transferred to a flask containing warm trypsin

(37°C). Then contents are stirred, and at an interval of every thirty minutes, the supernatant containing the dissociated cells can be collected. After removal of trypsin, the cells are dispersed in a suitable medium and preserved (by keeping the vial on ice).

The process of addition of fresh trypsin (to the tissue pieces), incubation and collection of dissociated cells (at 30 minutes intervals) is carried out for about 4 hours. The disaggregated cells are pooled, counted, appropriately diluted and then incubated.

Cold trypsinization This technique is more appropriately referred to as trypsinization with cold preexposure. The risk of damage to the cells by prolonged exposure to trypsin at 37°C (in warm trypsinization) can be minimized in this technique.

After chopping and washing, the tissue pieces are kept in a vial (on ice) and soaked with cold trypsin for about 6-24 hours. The trypsin is removed and discarded. However, the tissue pieces contain residual trypsin. These tissue pieces in a medium are incubated at 37°C for 20-30 minutes. The cells get dispersed by repeated pipettings. The dissociated cells can be counted, appropriately diluted and then used.

The cold trypsinization method usually results in a higher yield of viable cells with an improved survival of cells after 24 hours of incubation. This method does not involve stirring or centrifugation, and can be conveniently adopted in a laboratory. The major limitation of cold trypsinization is that it is not suitable for disaggregation of cells from large quantities of tissues.

Limitations of trypsin disaggregation Disaggregation by trypsin may damage some cells (e.g. epithelial cells) or it may be almost ineffective for certain tissues (e.g. fibrous connective tissue). Hence other enzymes are also in use for dissociation of cells.

Disaggregation by collagenase Collagen is the most abundant structural protein in higher animals. It is mainly present in the extra cellular matrix of connective tissue and muscle. The enzyme collagenase (usually crude one contaminated with non-specific proteases) can be effectively used for the disaggregation of several tissues (normal or malignant) that may be sensitive to trypsin. Highly purified grades of collagenase have been tried, but they are less effective when compared to crude collagenase.

The desired tissue suspended in basal salt solution, containing antibiotics is chopped into pieces. These pieces are washed by settling, and then suspended in a complete medium containing collagenase. After incubating for 1-5 days, the tissue pieces are dispersed by pipetting. The clusters of cells are separated by settling. The epithelial cells and fibroblastic cells can be separated.

Collagenase disaggregation has been successfully used for human brain, lung and several other epithelial tissues, besides various human tumors, and other animal tissues. Addition of another enzyme hyaluronidase (acts on carbohydrate residues on cell surfaces) promotes disaggregation. Collagenase in combination with hyaluronidase is found to be very effective for dissociating rat or rabbit liver. This can be done by perfusing the whole organ *in situ*. Some workers use collagenase in conjunction with trypsin, a formulation developed in chick serum, for disaggregation of certain tissues.

Use of other enzymes in disaggregation Trypsin and collagenase are the most widely used enzymes for disaggregation. Certain bacterial proteases (e.g. pronase, dispase) have been used with limited success. Besides hyaluronidase (described above), neuraminidase is also used in conjunction with collagenase for effective degradation of cell surface carbohydrates.

Primary explant technique

The primary explant technique was, in fact the original method, developed by Harrison in 1907. This technique has undergone several modifications, and is still in use. The tissue in basal salt solution is finely chopped, and washed by settlings. The basal salt solution is then removed. The tissue pieces are spread evenly over the growth surface. After addition of appropriate medium, incubation is carried out for 3-5 days. Then the medium is changed at weekly intervals until a substantial outgrowth of cells is observed. Now, the explants are removed and transferred to a fresh culture vessel.

The primary explant technique is particularly useful for disaggregation of small quantities of tissues (e.g. skin biopsies). The other two techniques - mechanical or enzymatic disaggregation, however, are not suitable for small amounts of tissues, as there is a risk of losing the cells. The limitation of explant technique is the poor adhesiveness of certain tissues to the growth surface, and the selection of cells in the outgrowth. It is however, observed that the primary explant technique can be used for a majority of embryonic cells e.g. fibroblasts, myoblasts, epithelial cells and glial cells.

Separation of viable and non-viable cells

It is a common practice to remove the nonviable cells while the primary culture is prepared from the disaggregated cells. This **is usually done when the first change of the medium is carried out**. The very few left over non-viable cells get diluted and gradually disappear as the proliferation of viable cells commences.

Sometimes, the non-viable cells from the primary cultures may be removed by centrifugation. The cells are mixed with ficoll and sodium metrizoate, and centrifuged. The dead cells form a pellet at the bottom of the tube.

Cell Lines

The term cell line refers to the **propagation of culture after the first subculture**. In other words, once the primary culture is subcultured, it becomes a cell line. A given cell line contains several cell lineages of either similar or distinct phenotypes.

It is possible to select a particular cell lineage by cloning or physical cell separation or some other selection method. Such a **cell line derived by selection or cloning is referred to as cell strain**. Cell strains do not have infinite life, ad they die after some divisions.

Finite cell lines

The cells in culture divide only a limited number of times, before their growth rate declines and they eventually die. The **cell lines with limited culture life spans are referred to as finite cell lines**. The cells normally divide 20 to 100 times (i.e. is 20-100 population doublings) before extinction. The actual number doubling depends on the species, cell lineage differences, conditions etc. Human cells generally divide 50-100 times, while murine cells divide 30-50 times before dying.

Continuous cell lines

A few cells in culture may acquire a different morphology and get altered. Such cells are capable of growing faster resulting in an independent culture. The progeny derived from these altered cells has unlimited life (unlike the cell strains from which they originated). They are designated as continuous cell lines. The **continuous cell lines are transformed, immortal and tumorigenic.** The transformed cells for continuous cell lines may be obtained from normal primary cell cultures (or cells strains) by treating them with chemical carcinogens or by infecting with oncogenic viruses.

Table 5 Comparison of Properties of Finite and Continuous Cell Lines

Property	*Finite cell line*	*Continuous cell line*
Growth rate	Slow	Fast
Mode of growth	Monolayer	Suspension or Monolayer
Yield	Low	High
Transformation	Normal	Immortal, tumorigenic
Ploidy	Euploid (multiple of haploid Chromosomes)	Aneuploid (not an exact multiple of haploid chromosomes)
Anchorage dependence	Yes	No
Contact inhibition	Yes	No
Cloning efficiency	Low	High
Serum requirement	High	Low
Markers	Tissue Specific	Chromosomal, antigenic or enzymatic

The most commonly used terms while dealing with cell lines are explained below:

Split ratio: The divisor of the dilution ratio of a cell culture at subculture. For instance, when each subculture divided the culture to half, the split ratio is 1:2.

Passage number: It is the number of times that the culture has been subcultured.

Generation number: It refers to **the number of doublings that a cell population has undergone.**

Nomenclature of cell lines

It is a common practice to give codes or designations to cell lines for their identification. For instance, the code **NHB 2-1** represents the cell line from normal human brain, followed by cell strain (or cell line number) **2** and clone number **1**. The usual practice in a culture laboratory is to maintain a logbook or computer database file for each of the cell lines.

While naming the cell lines, it is absolutely necessary to ensure that each cell line designation is unique so that there is no confusion when reports are given in literature. Further, at the time of publication, the cell line should be prefixed with a code designating the laboratory from which it was obtained, e.g., NCI for National Cancer Institute, WI for Wistar Institute.

Selection of cell lines

Several factors need to be considered while selecting a cell line. Some of them are briefly described.

1. Species: In general, non-human cell lines have less risk of biohazards, hence preferred. However, species differences need to be taken into account while extrapolating the data to humans.
2. Finite or continuous cell lines: Cultures with continuous cell lines are preferred as they grow faster, easy to clone and maintain, and produce higher yield. But it is doubtful whether the continuous cell lines express the right and appropriate functions of the cells. Therefore, some workers suggest the use of finite cell lines, although it is difficult.
3. Normal or transformed cells: The transformed cells are preferred as they are immortalized and grow rapidly.
4. Availability: The ready availability of cell lines is also important. Sometimes, it may be necessary to develop a particular cell line in a laboratory.

Table 6 A Selected List of Commonly used Cell Lines

Cell line	*Species of origin*	*Tissue of origin*	*Morphology*	*Ploidy*	*Characteristics*
IMR-90	Human	Lung	Fibroblast	Diploid	Susceptible to human viral infections.
3T3-A31	Mouse	Connective tissue	Fibroblast	Aneuploid	Contact inhibited, readily transformed
BHK21-C13	Hamster (Syrian)	Kidney	Fibroblast	Aneuploid	Readily transformable
CHO-K1	Chinese hamster	Ovary	Fibroblast	Diploid	Simple karyotype
NRK49F	Rat	Kidney	Fibroblast	Aneuploid	Induction of suspension growth by TGF-α,β
BRL 3A	Rat	Liver	Fibroblast	Diploid	Produces IGF-2
Vero	Monkey	Kidney	Fibroblast	Aneuploid	Viral substrate and assay
Hela-S_3	Human	Cervical carcinoma	Epithelial	Aneuploid	Rapid growth, high plating efficiency,
Sk/HEP-I	Human	Hepatoma	Endothelial	Aneuploid	Factor VIII
Caco-2	Human	Colo-rectal carcinoma	Epithelial	Aneuploid with polarized	Forms tight monolayer support
MCF-7	Human	Breast tumor (diffusion)	Epithelial	Aneuploid	Estrogen receptor positive
Friend	Mouse	Spleen	Suspension	Aneuploid	Hemoglobin, growth hormone,

5. Growth characteristics: The following growth parameters need to be considered.
   - Population doubling time
   - Ability to grow in suspension
   - Saturation density (yield per flask)
   - Cloning efficiency
6. Stability: The stability of cell line with particular reference to cloning, generation of adequate stock and storage are important.
7. Phenotypic expression: It is important that the cell lines possess cells with the right phenotypic expression.

7.8 MAINTENANCE OF CELL CULTURES

For the routine and good maintenance of cell lines in culture (primary culture or subculture) the examination of cell morphology and the periodic change of medium are very important.

Cell Morphology

The cells in the culture must be examined regularly to check the health status of the cells, the absence of contamination, and any other serious complication (toxins in medium, inadequate nutrients etc).

Replacement of Medium

Periodic change of the medium is required for the maintenance of cell lines in culture, whether the cells are proliferating or nor-proliferating. For the proliferating cells, the medium needs to be changed more frequently when compared to non-proliferating cells. The time interval between medium changes depends on the rate of cell growth and metabolism. For instance, for rapidly growing transformed cells (e.g. He La), the medium needs to be changed twice a week, while for slowly growing non-transformed cells (e.g. IMR-90), and the medium may be changed once a week. Further, for rapidly prolifering cells, the subculturing has to be done more frequently than for the slowly growing cells.

The following factors need to be considered for the replacement of the medium:

1. Cell concentration: The cultures with high cell concentration utilize the nutrients in the medium faster than those with low concentration; hence the medium is required to be changed more frequently for the former.
2. A decrease in pH: A fall in the pH of the medium is an indication for change of medium. Most of the cells can grow optimally at pH 7.0 and they almost stop growing when the pH falls to 6.5. A further drop in pH (between 6.5 and 6.0), the cells may lose their viability. The rate of fall in pH is generally estimated for each cell line with a chosen medium. If the fall is less than 0.1 pH units per day, there is no harm even if the medium is not immediately changed. But when the fall is 0.4 pH units per day, medium should be changed immediately.
3. Cell type: Embryonic cells, transformed cells and continuous cell lines grow rapidly and require more frequent subculturing and change of medium. This is in contract to normal cells, which grow slowly.
4. Morphological changes: Frequent examination of cell morphology is very important in culture techniques. Any deterioration in cell morphology may lead to an irreversible damage to cells. Change of the medium has to be done to completely avoid the risk of cell damage.

Subculture

Subculture (or passage) refers to the transfer of cells from one culture vessel to another culture vessel. Subculture usually (not always) involves the subdivision of proliferating cells that enables the propagation of a cell line. The term passage number is used to indicate the number of times a culture has been subculture.

The standard growth curve of cells in a culture includes an initial lag phase that is followed by exponential or log phase and a plateau phase. During the active growth period in the log phase, the medium must be changed frequently, or else growth ceases. As the cell concentration exceeds the capacity of the medium the culture has to be divided to subculture.

There are two types of subcultures–monolayer subculture and suspension subculture:

Monolayer Cultures

When the bottom of the culture vessel is covered with a continuous layer of cells, usually one cell in thickness, they are referred to as monolayer cultures. The attachment of cells among themselves and to the substrate (i.e. culture vessel) is mediated through surface glycoprotein's (cell adhesion molecules) and calcium ions (Ca^{2+}). For subculturing of monolayer cultures, it is usually necessary to remove the medium and dissociate the cells in the monolayer by degrading the cell adhesion molecules, besides removing Ca^{2+}.

Methods of cell dissociation

There are physical and enzymatic methods for dissociation of monolayer cultures. The commonly used cell dissociation methods for monolayer cultures:

Physical	
Mechanical shaking	Loosely adherent cells
Scraping	Cells sensitive to proteases
Enzymatic	
Trypsin	Most of the continuous cell lines
Trypsin + collagenase	Dense and multiplayer
Cultures Dispase	Removal of epithelium in sheets
Pronase	Single-cell suspensions

Scraping is employed for cultures, which are loosely adhered, and the use of proteases has to be avoided. Among the enzymes, trypsin is the most frequently used. For certain cell monolayer, which cannot be dissociated by trypsin, other enzymes such as pronase, dispase and collagenase are used. Prior to cell dissociation by enzymes, the monolayers are usually subjected to pretreatment of EDTA for the removal of Ca^{2+}.

Criteria for subculture of monolayers

The subculturing is ideally carried out between the middle of the log phase and the time before they enter plateau phase. Subculture of cells should not be done when they are in lag phase. The other important criteria for subculture of monolayers are briefly described.

Culture density

It is advisable to subculture the normal or transformed monolayer cultures, as soon as they reach confluence. Confluence denotes the culture stage wherein all the available growth area is utilized and the cells make close contact with each other.

Medium exhaustion

A drop in PH is usually accompanied by an increase in culture cell density. Thus, when the pH falls, the medium must be changed, followed by subculture.

Scheduled timing of subculture

It is now possible to have specified schedule timings for subculture of each cell line. For a majority of cell cultures, the medium change is usually done after 3-4 days, and subculturing after 7 days.

Purpose of subculture

The purpose for which the cells are required is another important criterion for consideration of subculturing. Generally, if the cells are to be used for any specialized purpose, they have to be subculture more frequently.

Techniques of monolayer subculture

The subculture of monolayer cells basically consists of the following steps:

1. Removal of the medium
2. Brief exposure of the cells to trypsin

3. Removal of trypsin and dispersion in a medium
4. Incubation of cells to round up
5. Resuspension of the cells in a medium for counting and reseeding
6. Cells reseeded and grown to monolayers

Cell concentration at subculture

Most of the continuous cell lines are subculture at a seeding concentration between 1×10^4 and 5×10^4 cells/ml. However for a new culture, subculture has to be started at a high concentration and gradually reduced.

Suspension Cultures

Majority of continuous cell lines grow as monolayer. Some of the cells which are nonadhesive e.g. cells of leukemia or certain cells which can be mechanically kept in suspension, can be propagated in suspension. The transformed cells are subcultured by this method. Subculture by suspension is comparable to culturing of bacteria or yeast.

Advantages of cell propagation by suspension

- The process of propagation is faster
- The lag period is usually shorter
- Results in homogeneous suspension of cells
- Treatment with trypsin is not required
- Scale-up is convenient
- No need for frequent replacement of the medium
- Maintenance is easy
- Bulk production of cells can be conveniently achieved

Criteria for suspension subculture

The criteria adopted for suspension subculture are the same as that already described for monolayer subcultures. The following aspects have to be considered:

- Culture density
- PH change representing medium exhaustion
- Schedule timings of subculture
- Purpose of subculture

Technique of suspension culture

The cells can be suspended in a culture flask (a stirrer flask) containing the desired medium. The medium is continuously stirred with a magnetic pendulum rotating at the base of the flask. The cells have to periodically examine for contamination or signs of deterioration.

Stem Cell Cultures

The cells that retain their proliferate capacity throughout life are regarded as stem cells. When the stem cells divide, they can generate differentiated cells and/or some more stem cells. These stem cells are capable of regenerating tissue after injury. The lack of tissue-specific differentiation markers is a characteristic feature of stem cells.

Embryonic Stem (ES) Cells

As the embryonic development occurs, cells of the **inner mass of embryo** (i.e. those contributing to future fetus) represent embryonic stem (ES) cells. They continue to divide and remain in an undifferentiated to totipotent state. It has been possible to establish and maintain cell lines for ES cells. The ES cells isolated from mouse blastocyst are the most commonly used in the laboratory. The most widely used embryonic stem lines are the various 3T3 lines, WI-38, MRC-5 and other human fetal lung fibroblasts.

Advantage of ES cells

In general, the cultures from embryonic tissues survive, and proliferate better than those from the adults. This is due to the fact that ES cells are less specialized with higher proliferative potential.

Limitations of ES cells

In some cases, the ES cells will be different from the adult cells, and thus there is no guarantee that they will mature to adult-type cells. Therefore, it is necessary to characterize the cells by appropriate methods.

Epithelial Stem Cells

The epithelial cells (e.g., epidermis lining of gut) are constantly being shed from their outer surface. This cellular loss is compensated by a continuous replacement progress. The replacement occurs in a highly organized and a regulated fashion.

It is estimated the in humans the entire outer layer is skin is shed daily. The entire epithelial lining of the mouse gut is replaced once in 3-4 days. This process of shedding and replacement continues throughout life. The epidermis of the skin has a proliferative compartment containing stem cells and postmitotic cells, besides some transit amplifying population of cells. The transit amplifying cells, produced from cells with limited life span are shed from the epidermis.

Maintenance of Stem Cells in Culture

The basic criteria to maintain stem cell *in vitro* is to ensure that they possess the same characteristics and differentiating abilities when they present in the tissue *in vivo.* The maintenance of epidermal and non-epidermal epithelial cells in the *in vitro* cultures is briefly described.

Epidermal stem cells in culture

The epidermal stem (or keratinocyte stem) cells can be successfully maintained by co-culturing with 3T3 feeder layer. By this technique, it is possible to achieve long-term maintenance of cells, besides retaining their capacity for both proliferative and differentiating characteristics. It has been demonstrated that the so maintained stem cells when placed in to nude mice could form stratified and differentiated epithelium. Serum-free media with added growth factors were found to be more efficient in maintaining the epidermal stem cells in culture.

Epithelial cells in culture

Several types of non-epidermal epithelial cells can be grown and maintained in cultures. As in the case with epidermal stem cells, use of feeder layer is advantageous for epithelial cell culture. Epithelial cells of prostate gland have been successful grown in suspension culture in the presence of 3T3 feeder layer. The same method is also used for culturing human breast epithelial cells and colorectal carcinoma cells.

Characterization of Stem Cells

Immunological techniques are widely used for the characterization of different populations of stem cells. These techniques are mostly based on immunocytochemistry using fluorescent microscopy or staining technique involving color reactions.

The cells of the tissues produce specific cell surface and cytoplasmic proteins. The cell surface proteins such as integrins and the members of CD (cluster of differentiation) antigens (e.g. CD10' CD31' CD44) can be used as markers of epithelial cell types. Further, the cytoplasmic proteins of epithelial cells (of cytokeratin family) are also useful for their identification.

Applications of Cultured Stem Cells

Embryonic stem cells in tissue repair

The culture stem cells can be used for the repair of tissues with functional impairment that may occur due to damage or ageing. The cultured embryonic stem cells can be manipulated to produce cultures characteristic of a particular tissue. Thus, there exists a possibility of treating the following diseases.

- Diabetes with pancreatic insulin producing cells
- Parkinson's disease with cultured dopamine-producing neurons

Embryonic stem cells are useful for the production of defined transgenic animals. It is also possible to modify ES cell genome by gene targeting using *in vitro* transformation and selection.

Applications of tissue specific stem cells

Stem cells, isolated from different tissues of humans and animals, and cultured *in vitro* are less totipotent than ES cells. They usually differentiate into a single cell type and are referred to as *unipotent*. However, stems cells from none marrow and brain though to a lesser extent when compared to ES cells. In mouse lacking bone marrow, when the cultured neuronal cells are placed, they develop into blood cells.

7.9 CELL VIABILITY AND CYTOTOXICITY

As the cell are removed from the living *(in vivo)* environment and subjected to experimental manipulations in the culture systems *(in vitro),* their viability assumes significance. Viability of the cells represents the capability of their existence, survival and development.

Many experiments are carried out with cells in the culture rather than using the animal models. This is particularly so with regard to the determination of safely and cytotoxicity of several compounds (pharmaceuticals, cosmetics, anticancer drugs, food additives) In vitro testing for cytotoxicity and safety evaluation is in fact cost-effective, besides reducing the use of animals. Studies on cytotoxicity broadly involve the metabolic alterations of the cells, including the death of cells as a result of toxic effects of the compounds. For instance, in case of anticancer drugs, one may look for death of cells, while for cosmetics the metabolic alterations and allergic responses may be more important.

Assays for Cell Viability and Cytotoxicity

Despite the limitation stated above, there are several assays developed in the laboratory for measuring the cell viability and cytotoxicity. They are broadly categorized into the following types:

- Cytotoxicity assays

- Survival assays
- Metabolic assays
- Transformation assays
- Inflammation assays

Cell Cloning

In the traditional culture techniques, the cells are heterogeneous in nature. Isolation of pure cell strains is often required for various purposes. *Cell cloning broadly involves* the process connected with the *production of a population of cells derived from a single cell.* Cloning of continuous cell lines is much easier compared to that of the primary cultures, and finite cell lines.

There are certain limitations for cloning of culture cells derived from normal tissues. These cells survive for a limited number of generations and therefore cloning may not result in any significant number of cells. On the other hand cloning of continuous cell lines due to their transformed status is much easier. Thus, the transformed cells have higher cloning efficiency compared to normal cells. Cloning may be carried our by two approaches monolayer and suspension cultures.

1. Monolayer culture

Petri dishes and multi-well plates or flasks can be used for cloning by monolayer culture. It is relatively easy to remove the individual colonies of cells from the surface where they are attached.

2. Suspension culture

Cloning can be carried out in suspension by seeding cells into viscous solutions or gels (agar). As the daughter cells are formed in suspension, they remain intact and form colonies in suspension.

Organ Cultures

The use of organ cultures (organ or their representative fragments) with reference to structural integrity, nutrient and gas exchange, growth and differentiation, along with the advantages and limitation is briefly described.

Structural integrity

As already stated, the isolated cells are individual, while in the organ culture, the cells are integrated as a single unit. The cell-to-cell association, and interactions found in the native tissue or organs are retained to a large extent.

As the structural integrity of the original tissue is preserved, the associated cells can exchange signals through cell adhesion or communications.

Nutrient and gas exchange

There is *no vascular system* in the organ culture. This *limits the nutrient supply and gas exchanges* of the cells. This happens despite the adequate care taken in the laboratory for the rapid diffusion of nutrients and gas by placing the organ cultures at the interface between the liquid and gaseous phases. As a consequence, some degree of necrosis at the central part of the organ may occur.

Some workers prefer to use high O_2 concentration (sometimes even pure O_2) in the organ cultures. Exposure of cells to high O_2 content is associated with the risk of O_2 induced toxicity e.g. nutrient metabolite exchange is severely affected.

Growth and differentiation

In general, the organ cultures do not grow except some amount of proliferation that may occur on the outer cell layers.

Advantage of organ cultures

- Provide a direct means of studying the behavior of an integrated tissue in the laboratory
- Understanding of biochemical and molecular functions of an organ/issue becomes easy

Limitation of organ cultures

- Organ cultures cannot be propagated, hence for each experiment there is a need for a fresh organ from a donor
- Variations are high and reproducibility is low
- Difficult to prepare, besides being expensive

Techniques of Organ Culture

The most important requirement of organ or tissue culture is to place them at such a location so that optimal nutrient and gas exchanges occur. This is mostly achieved by keeping the tissue at gas-limited interface of the following supports.

- Semisolid gel of agar
- Clotted plasma
- Microporous filter
- Les paper
- Strip of Perspex or Plexiglas

In recent years, ***filter-well inserts*** are in use to attain the natural geometry of tissues more easily.

Procedure for organ culture

The basic technique of organ culture consists of the following stages.

1. Dissection and collection of the organ tissue
2. Reduce the size of the tissue as desired, preferably to less than mm in thickness
3. Place tissue one support (listed above) at the gas medium interface
4. Incubate in humid CO_2 incubator
5. Change the medium (M199 or CMRL 1066) as frequently as desired
6. The organ culture can be analyzed by histology, autoradiography and immunochemistry.

Questions

1. Define the term Cell Line with suitable example.
2. Comment on the importance of the growth media in cell culture
3. List out the infrastructure facilities required for efficient plant and animal cell culture in laboratories.

4. Write a detailed note on the two common procedures practiced in eliminating the contamination in cell culture laboratories.
5. Explain the importance of Serum Free Media with regard to cell lines.
6. Write a brief essay on the distinguished characteristic features of cultured cells and their metabolic significance.
7. Give an elaborate view on the identification markers of cell line.
8. Explain in detail the methods adopted to identify specific cell lines.
9. List out the experiments that are required to give information about the measurement of cultured cells.
10. Name the method to analyze the cell proliferation and survival of cultured cells and explain the procedure in detail.
11. List out the applications and products of animal cell culture.
12. Give an elaborate sketch of the biohazards involved
13. Write a detailed essay on cell or cellular senescence
14. Comment on Apoptosis and the tests involved to identify an apoptotic cell.
15. Write a detailed note on finite and continuous cell line with suitable examples.
16. Sketch out the importance of maintaining cell culture
17. Write a detailed note on stem cells and its importance
18. Write a note on the assays of cell viability and cytotoxicity.
19. Give the importance of organ culture.
20. Write a brief note on cell cloning.

Index

About the Author

Dr Shaleesha A. Stanley is currently Professor and Head, Department of Biotechnology, Jeppiaar Engineering College, Chennai. With double M.Sc. and Ph.D from the Madurai Kamaraj University and University of Madras, she spent 13 years in research and teaching aquatic biotechnology and biological sciences. She has been associated with many educational innovations at the M.S. Swaminathan Research Foundation, Chennai. She has published several research papers in various journals of international repute.